P9-BZF-408

ALCOHOL

And Alcoholism

THE ENCYCLOPEDIA OF PSYCHOACTIVE DRUGS

IN 25 VOLUMES
Each title on a specific drug or drug-related problem

ALCOHOL

And Alcoholism

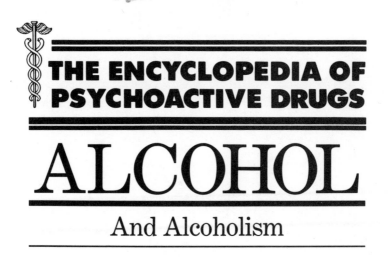

THE ENCYCLOPEDIA OF PSYCHOACTIVE DRUGS

ALCOHOL

And Alcoholism

ROSS FISHMAN, Ph.D.

CHELSEA HOUSE PUBLISHERS
NEW YORK
NEW HAVEN   PHILADELPHIA

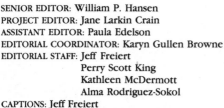

SENIOR EDITOR: William P. Hansen
PROJECT EDITOR: Jane Larkin Crain
ASSISTANT EDITOR: Paula Edelson
EDITORIAL COORDINATOR: Karyn Gullen Browne
EDITORIAL STAFF: Jeff Freiert
        Perry Scott King
        Kathleen McDermott
        Alma Rodriguez-Sokol
CAPTIONS: Jeff Freiert
ART COORDINATOR: Carol McDougall
LAYOUT: Victoria Tomaselli
ART ASSISTANT: Noreen M. Lamb
PICTURE RESEARCH: Elizabeth Terhune
        Tamara Fulop
        Diane Wallis

COVER: Stockphotos, Inc./Henry Wolf

   3 5 7 9 8 6 4

Library of Congress Cataloging in Publication Data
Fishman, Ross.

  Alcohol and alcoholism.
    (The Encyclopedia of psychoactive drugs)
    Bibliography: p.
    Includes index.
    Summary: Examines alcohol abuse and alcoholism, a
    disease many feel cannot be cured, only arrested.
      1. Alcoholism—Juvenile literature. 2. Alcohol—
Physiological effect—Juvenile literature.
[1. Alcoholism. 2. Alcohol] I. Title. II. Series.
RC565.F557 1986   616.86′1   85-32572
ISBN 0-87754-762-9

CONTENTS

Three men enjoy drinks at an outdoor cafe. Although many Americans drink for pleasure and in moderation, alcohol is a powerful drug that can bring incalculable misery not only to those who abuse it, but also to the friends and relatives of problem drinkers.

FOREWORD

In the Mainstream of American Life

The rapid growth of drug use and abuse is one of the most dramatic changes in the fabric of American society in the last 20 years. The United States has the highest level of psychoactive drug use of any industrialized society. It is 10 to 30 times greater than it was 20 years ago.

According to a recent Gallup poll, young people consider drugs the leading problem that they face. One of the legacies of the social upheaval of the 1960s is that psychoactive drugs have become part of the mainstream of American life. Schools, homes, and communities cannot be "drug proofed." There is a demand for drugs—and the supply is plentiful. Social norms have changed and drugs are not only available—they are everywhere.

Almost all drug use begins in the preteen and teenage years. These years are few in the total life cycle, but critical in the maturation process. During these years adolescents face the difficult tasks of discovering their identity, clarifying their sexual roles, asserting their independence, learning to cope with authority, and searching for goals that will give their lives meaning. During this intense period of growth, conflict is inevitable and the temptation to use drugs is great. Drugs are readily available, adolescents are curious and vulnerable, there is peer pressure to experiment, and there is the temptation to escape from conflicts.

No matter what their age or socioeconomic status, no group is immune to the allure and effects of psychoactive drugs. The U.S. Surgeon General's report, "Healthy People," indicates that 30% of all deaths in the United States

Young people socialize at a Washington, D.C., bar. Although alcohol often serves as an efficient tension reliever in social situations, many teenagers and young adults come to rely on its effects, a habit that often leads to addiction and other problems.

are premature because of alcohol and tobacco use. However, the most shocking development in this report is that mortality in the age group between 15 and 24 has increased since 1960 despite the fact that death rates for all other age groups have declined in the 20th century. Accidents, suicides, and homicides are the leading cause of death in young people 15 to 24 years of age. In many cases the deaths are directly related to drug use.

THE ENCYCLOPEDIA OF PSYCHOACTIVE DRUGS answers the questions that young people are likely to ask about drugs, as well as those they might not think to ask, but should. Topics include: what it means to be intoxicated; how drugs affect mood; why people take drugs; who takes them; when they take them; and how much they take. They will learn what happens to a drug when it enters the body. They will learn what it means to get "hooked" and how it happens. They will learn how drugs affect their driving, their schoolwork, and those around them—their peers, their family, their friends, and their employers. They will learn what the signs are that indicate that a friend or a family member may have a drug problem and to identify four stages leading from drug use to drug abuse. Myths about drugs are dispelled.

National surveys indicate that students are eager for information about drugs and that they respond to it. Students not only need information about drugs—they want information. How they get it often proves crucial. Providing young people with accurate knowledge about drugs is one of the most critical aspects.

THE ENCYCLOPEDIA OF PSYCHOACTIVE DRUGS synthesizes the wealth of new information in this field and demystifies this complex and important subject. Each volume in the series is written by an expert in the field. Handsomely illustrated, this multi-volume series is geared for teenage readers. Young people will read these books, share them, talk about them, and make more informed decisions because of them.

Miriam Cohen., Ph.D.
Contributing Editor

Alcohol is the drug most widely used by Americans under 25. Although young people tend to drink less regularly than older people, when they do drink they usually consume greater quantities.

INTRODUCTION

The Gift of Wizardry
Use and Abuse

JACK H. MENDELSON, M.D.
NANCY K. MELLO, PH.D.
Alcohol and Drug Abuse Research Center
Harvard Medical School—McLean Hospital

Dorothy to the Wizard:

"I think you are a very bad man," said Dorothy.
"Oh, no, my dear; I'm really a very good man; but I'm a very bad
Wizard."

—from THE WIZARD OF OZ

Man is endowed with the gift of wizardry, a talent for discovery and invention. The discovery and invention of substances that change the way we feel and behave are among
man's special accomplishments, and like so many other products of our wizardry, these substances have the capacity to
harm as well as to help. The substance itself is neutral, an
intricate molecular structure. Yet, "too much" can be sick-
ening, even deadly. It is man who decides how each substance
is used, and it is man's beliefs and perceptions that give this
neutral substance the attributes to heal or destroy.

Consider alcohol—available to all and yet regarded with
intense ambivalence from biblical times to the present day.
The use of alcoholic beverages dates back to our earliest
ancestors. Alcohol use and misuse became associated with
the worship of gods and demons. One of the most powerful
Greek gods was Dionysus, lord of fruitfulness and god of wine.
The Romans adopted Dionysus but changed his name to Bacchus. Festivals and holidays associated with Bacchus cele-
brated the harvest and the origins of life. Time has blurred
the images of the Bacchanalian festival, but the theme of
drunkenness as a major part of celebration has survived the
pagan gods and remains a familiar part of modern society.

The term "Bacchanalian festival" conveys a more appealing image than "drunken orgy" or "pot party," but whatever the label, some of the celebrants will inevitably start up the "high" escalator to the next plateau. Once there, the de-escalation is difficult for many.

According to reliable estimates, one out of every ten Americans develops a serious alcohol-related problem sometime in his or her lifetime. In addition, automobile accidents caused by drunken drivers claim the lives of tens of thousands every year. Many of the victims are gifted young people, just starting out in adult life. Hospital emergency rooms abound with patients seeking help for alcohol-related injuries.

Who is to blame? Can we blame the many manufacturers who produce such an amazing variety of alcoholic beverages? Should we blame the educators who fail to explain the perils of intoxication, or so exaggerate the dangers of drinking that no one could possibly believe them? Are friends to blame— those peers who urge others to "drink more and faster," or the macho types who stress the importance of being able to "hold your liquor"? Casting blame, however, is hardly constructive, and pointing the finger is a fruitless way to deal with problems. Alcoholism and drug abuse have few culprits but many victims. Accountability begins with each of us, every time we choose to use or to misuse an intoxicating substance.

It is ironic that some of man's earliest medicines, derived from natural plant products, are used today to poison and to intoxicate. Relief from pain and suffering is one of society's many continuing goals. Over 3,000 years ago, the Therapeutic Papyrus of Thebes, one of our earliest written records, gave instructions for the use of opium in the treatment of pain. Opium, in the form of its major derivative, morphine, remains one of the most powerful drugs we have for pain relief. But opium, morphine, and similar compounds, such as heroin, have also been used by many to induce changes in mood and feeling. Another example of man's misuse of a natural substance is the coca leaf, which for centuries was used by the Indians of Peru to reduce fatigue and hunger. Its modern derivative, cocaine, has important medical use as a local anesthetic. Unfortunately, its increasing abuse in the 1980s has reached epidemic proportions.

The purpose of this series is to provide information about the nature and behavioral effects of alcohol and drugs, and the probable consequences of their use. The information presented here (and in other books in this series) is based on many clinical and laboratory studies and observations by people from diverse walks of life.

Over the centuries, novelists, poets, and dramatists have provided us with many insights into the beneficial and problematic aspects of alcohol and drug use. Physicians, lawyers, biologists, psychologists, and social scientists have contributed to a better understanding of the causes and consequences of using these substances. The authors in this series have attempted to gather and condense all the latest information about drug use and abuse. They have also described the sometimes wide gaps in our knowledge and have suggested some new ways to answer many difficult questions.

One such question, for example, is how do alcohol and drug problems get started? And what is the best way to treat them when they do? Not too many years ago, alcoholics and drug abusers were regarded as evil, immoral, or both. It is now recognized that these persons suffer from very complicated diseases involving complex biological, psychological, and social problems. To understand how the disease begins and progresses, it is necessary to understand the nature of the substance, the behavior and genetic makeup of the afflicted person, and the characteristics of the society or culture in which he lives.

The diagram below shows the interaction of these three factors. The arrows indicate that the substance not only affects the user personally, but the society as well. Society influences attitudes towards the substance, which in turn affect its availability. The substance's impact upon the society may support or discourage the use and abuse of that substance.

SUBSTANCE
(ALCOHOL OR DRUG)

PERSON ↔ SOCIETY

An intoxicated bar patron. There are presently more than 10 million people with drinking problems in the United States, and the percentage of women in this category has increased dramatically.

Although many of the social environments we live in are very similar, some of the most subtle differences can strongly influence our thinking and behavior. Where we live, go to school and work, whom we discuss things with—all influence our opinions about drug use and misuse. Yet we also share certain commonly accepted beliefs that outweigh any differences in our attitudes. The authors in this series have tried to identify and discuss the central, most crucial issues concerning drug use and misuse.

Regrettably, man's wizardry in developing new substances in medical therapeutics has not always been paralleled by intelligent usage. Although we do know a great deal about the effects of alcohol and drugs, we have yet to learn how to impart that knowledge, especially to young adults.

Does it matter? What harm does it do to smoke a little pot or have a few beers? What is it like to be intoxicated? How long does it last? Will it make me feel really fine? Will it make me sick? What are the risks? These are but a few of the questions answered in this series, which, hopefully, will enable the reader to make wise decisions concerning the crucial issue of drugs.

Information sensibly acted upon can go a long way towards helping everyone develop his or her best self. As one keen and sensitive observer, Dr. Lewis Thomas, has said,

> There is nothing at all absurd about the human condition. We matter. It seems to me a good guess, hazarded by a good many people who have thought about it, that we may be engaged in the formation of something like a mind for the life of this planet. If this is so, we are still at the most primitive stage, still fumbling with language and thinking, but infinitely capacitated for the future. Looked at this way, it is remarkable that we've come as far as we have in so short a period, really no time at all as geologists measure time. We are the newest, the youngest, and the brightest thing around.

Two men on a company picnic pose in back of a trash can filled with empty beer cans. Even during the Prohibition era, drinking was an undeniable part of American society. Today, an estimated 175 million Americans consume some form of alcohol.

AUTHOR'S PREFACE

There is no denying that the United States is a drinking society. In fact, alcohol has been with us since the day the first settlers from Europe arrived on these shores. Today nearly every town and city in the United States has an abundance of places where alcohol can be purchased and/or consumed: bars and restaurants, liquor stores, retail food stores, private clubs, and even many volunteer fire departments.

Unfortunately, of the 175 million Americans who consume alcohol, approximately 10–13 million suffer from alcoholism. In addition, over half of all hospital emergency-room admissions involve alcohol-related accidents or illnesses. Aside from health factors, the large consumption of alcohol affects American society on many levels. The production, distribution, and marketing of alcoholic beverages employs many people and generates huge sums of money. Also, federal and state taxes on alcohol produce enormous revenues that contribute significantly to the economy. Clearly, because drinking is such an integral part of our lives, the problems associated with alcohol abuse are not about to disappear quickly.

This book will explore a variety of subjects: society's changing attitudes toward alcohol and alcoholism from antiquity to the present; the physiological and psychological consequences of alcoholism; the high cost of alcoholism to the victim, his or her family, and society; and the various methods of treatment currently available to fight this widespread disease. Finally, the book will discuss recent trends in alcohol prevention and education to see the ways in which experts in the field are attempting to combat alcoholism before it begins its progress through individual lives.

This poster depicts the 1873–74 women's crusade against the sale, use, and abuse of alcoholic beverages. The crusade led to the establishment of a national organization, the Women's Christian Temperance Union (WCTU), in 1874, which supported prohibition laws.

CHAPTER 1

THE HISTORY OF ALCOHOL USE

Should any doubt exist that alcoholic beverages have been in existence for a long time, one need only turn to the Bible, where many references to alcohol and its use may be found. For example, according to the book of Genesis, after Noah survived the Great Flood he began a new livelihood: growing grapes and making wine, which he frequently drank to the point of intoxication.

Although the Bible clearly illustrates the negative consequences of excessive drinking, alcohol in the form of wine was seen as a gift of God to mankind. Drunkenness, however, was widely disparaged because, as is claimed in the biblical Proverbs, it leads to arguments, poverty, and madness. Clearly, then, the enjoyment and abuse of alcohol have existed side by side for centuries.

Early Attitudes in the United States

The view of alcohol as a beneficial gift of God also prevailed in colonial America, where the use of alcoholic beverages was widely accepted. Alcohol was considered a medicinal healer for the sick, a tonic for the healthy, an uplifter for the saddened, and required fare for celebrants. Only its abuse was condemned, and drunkards, as abusers of God's gift, were seen as sinners.

As an increasing number of breweries and stills turned out greater amounts of alcoholic beverages at decreasing prices, drunkenness became a growing problem. In 1632, after an early attempt to reduce excessive use of alcohol failed, the Virginia colony outlawed drunkenness. Public demand for more drinking places, however, brought about a steady increase of licensed taverns and inns, and eventually there were just too many taverns to monitor. Early in the 18th century, growing concern that the abuse of alcoholic beverages was getting out of hand led to attempts to establish a prohibition on all alcohol. In Georgia a nine-year effort was made to prevent the importation of rum, and in Virginia there was a similar attempt at restriction. Both attempts failed.

Following the American Revolution, alcohol consumption increased dramatically — one of many changes in this freer post-revolutionary period. In addition to a shift in preference from beer and wine to spirits such as rum and whiskey, drinking patterns themselves began to undergo gradual change. Now, rather than just being a part of socializing or celebrating, alcohol consumption was becoming more of an end in itself. The custom of the family drinking together, still popular in England, slowly gave way to the male-centered

The recorded use of alcohol dates back to Biblical times. This illustration depicts the sons of Noah watching their sleeping father, who, according to the book of Genesis, cultivated a vineyard, became drunk on wine, and fell asleep.

THE BETTMANN ARCHIVE

tavern. Contributing to these changes were increased production of alcoholic beverages and a general increase in the population, which led to more crowded cities and greater westward migration. With alcohol prices decreasing and more taverns being established, people from the lower and middle classes had greater access to liquor.

Temperance and Prohibition

In the 1820s, a movement began in earnest in the United States to encourage temperance (minimal or moderate use of alcohol). By 1832 most states had at least one temperance organization operating within them. Advocates of temperance first relied on persuasion as the way to bring about moderation. When this failed, however, the temperance reformers attempted to create regulations for the licensing of taverns, and urged the taxation of alcoholic beverages. Quite contrary to expectation, the rapid growth of the temperance movement in the 1830s was paralleled by a dramatic increase in alcohol consumption — to a level that has not been surpassed since.

This 19th-century woodcut graphically depicts the alcoholic as a wife beater and child abuser, a common stereotype at that time.

Over the next 20 years, however, consumption dropped, and the temperance movement continued to gain strength. Using political action, the movement succeeded in establishing prohibition laws in 13 states. By 1865, though, these laws had all been repealed.

During the 1880s and 1890s, as anti-alcohol sentiment grew in fervor, many states passed or reenacted prohibition laws. Although by the beginning of the 20th century these sentiments had again waned, they came back with renewed force as the southwestern states turned to prohibition. In 1919 the U.S. Congress adopted the 18th Amendment, which prohibited the manufacture, importation, exportation, transportation, and sale of alcoholic beverages in the United States and its possessions.

Prohibition went into effect in 1920, but from the outset it had at least two serious handicaps. First, it was nearly impossible to enforce the law's provisions. Second, the law did not prohibit the *purchase* or *use* of alcoholic beverages. In fact, vast fortunes were amassed during Prohibition by entrepreneurs involved in the production and sale of "bootlegged," or illegal, alcohol.

An official during the 1920s destroys a barrel of beer. Prohibition, which went into effect in 1920, prohibited the manufacture, importation, exportation, transportation, and sale of alcohol in the United States.

THE BETTMANN ARCHIVE

By the early 1930s pressure to end Prohibition had grown; indeed, during the 1932 national political conventions both the Democratic and Republican parties advocated repeal. The 21st Amendment to the Constitution, passed shortly after Franklin Roosevelt became president in 1933, once again made liquor legal.

Those people who favored Prohibition showed no great concern for those who suffered from overindulgence in alcohol. Drunkenness was still viewed by many of Prohibition's supporters as sinful behavior, and relapses were regarded as further evidence of moral weakness. Therefore, although during the Prohibition era the ravages of alcoholism were increasing and drunk driving was on the rise, little help was available to people wishing to overcome their dependence on alcohol. Alcoholism was not yet considered a medical disorder and disease, and thus was not the responsibility of health professionals.

A float in a New York City anti-Prohibition parade. In 1933 public pressure finally resulted in the repeal of this law.

A liquor store owner posts a warning to drunk drivers. Concern over the rise in alcohol-related highway fatalities has led to harsher penalties for those convicted of driving while intoxicated.

CHAPTER 2

CHANGING ATTITUDES TOWARDS ALCOHOLISM

*T*he first major breakthrough in the evolution of enlightened attitudes toward alcoholism took place in 1935, when Bill Wilson, a stockbroker and chronic alcoholic, had what he described as a mystical experience that resulted in his being able to stop drinking. He shared this experience with a certain Dr. Bob, a physician who was also an alcoholic. Working together, the two men devised a nonprofessional self-help group.

Known as Alcoholics Anonymous (AA), the group emphasized mutual support, a commitment to abstinence, anonymity, and a 12-step program toward recovery from alcoholism. AA grew steadily, spreading across the United States and around the world. The organization has served as a model for numerous other self-help groups and has helped thousands of individuals from all walks of life recover from alcoholism.

Modern Treatment Patterns

Despite the rapid growth and apparent success of Alcoholics Anonymous over the past 50 years, society in general, and the health care sector in particular, have paid relatively little attention to alcohol-related problems. Because in earlier decades drinking remained generally acceptable behavior, alcohol abusers and alcoholics were typically viewed as an unfortunate minority of weak-willed people, victims of self-inflicted damage and hardship.

In 1956, however, in a second major breakthrough that followed by more than 20 years the establishment of Alcoholics Anonymous, the American Medical Association recognized alcoholism as a disease. But during the 1960s and 1970s a growing fear of marijuana, heroin, and prescription-drug abuse largely overshadowed concern over alcoholism. Government financing of a massive effort to eradicate drug abuse from our society took precedence over any attempts to address the more familiar — and therefore more tolerated — problem of alcoholism. Some sociologists and epidemiologists did maintain a keen interest in the patterns of alcohol use and abuse across the country. Undoubtedly, their interest was fueled by the fact that by the 1960s there were an estimated 5 million alcoholics in the United States.

Spurred by the persistent efforts of Iowa Senator Harold Hughes, himself a recovered alcoholic, and his supporters to gain government funding for research on alcoholism, Congress passed the Hughes Bill of 1970, which established the National Institute of Alcohol Abuse and Alcoholism (NIAAA). The law's official recognition of alcoholism as a disease with

In 1935 Dr. Bob, a physician and alcoholic, co-founded Alcoholics Anonymous, a self-help group that stresses frequent meetings. AA's members first admit that their drinking has made their lives unmanageable, and then make a commitment to abstinence.

ALCOHOLICS ANONYMOUS

grave implications for the country finally raised the issue to a level of importance already occupied by mental health and drug abuse issues.

Soon after its conception, the NIAAA commissioned a study of the extent of alcoholism in the United States. The study estimated there were 9 million alcoholics in the country. (The number now is thought to be between 10 and 13 million.) The NIAAA then began to tally the costs that alcohol abuse and alcoholism impose on society: loss of life; work accidents and injuries; injuries and deaths related to drunk driving; loss of productivity at work due to lateness, absenteeism, and poor performance; the burden of maintaining direct health and mental health care services for alcoholics and their families; and the break up of families.

To help reduce these costs, the NIAAA began, both directly and through state alcoholism authorities, to fund (1) the establishment of treatment programs, (2) a program for educating the general public about alcoholism, (3) research into questions about alcoholism, and (4) the development of programs to prevent alcohol abuse. Special attempts were made to raise alcohol-related issues among people thought to be most vulnerable to alcoholism or people likely to suffer the most from its effects: teenagers, the elderly, women, blacks, Hispanics, Native Americans, the poor, people with disabilities, and people who are dependent upon other substances in addition to alcohol.

As a result of the growing awareness that chronic excessive alcohol consumption is a disease with serious consequences not only for the alcohol abuser but for society as a whole, a new set of attitudes developed towards drinking and alcohol-induced behavior. Nowhere is this more evident than in the recent protest against drunk driving and the more severe penalties now being imposed upon those people found guilty of driving while intoxicated.

Half of all automobile accidents, the leading cause of death among teenagers, are alcohol-related. Until recently, drunk drivers who caused deaths and injuries could usually count on receiving light punishments, or even no punishments at all. A few years ago, however, a mother whose child was killed by a drunk driver started a movement called Mothers Against Drunk Drivers (MADD). MADD complained that a suspended license or a few months in jail is nothing more than "a slap on the wrist" for destroying property or causing loss of life. The group has grown rapidly into a well-financed national organization that educates the public. As a result of pressure from MADD, various states have adopted new penalties for those convicted of drunk driving. These include the automatic suspension of the driver's license, mandatory attendance of educational classes, and increases in car insurance premiums.

Young people, who often are both the cause and the victims of alcohol-related accidents, have taken up the cause through Students Against Drunk Drivers (SADD) and other similar organizations, which operate in many communities and in high schools and college campuses across the country. One popular program, Safe Rides, offers a hotline to young party-goers who need to find safe transportation home after they or the person who drove them to the party have become intoxicated.

In recent years the public has become more aware of the fact that alcohol abuse contributes significantly to the reduced efficiency of workers, resulting in lower productivity

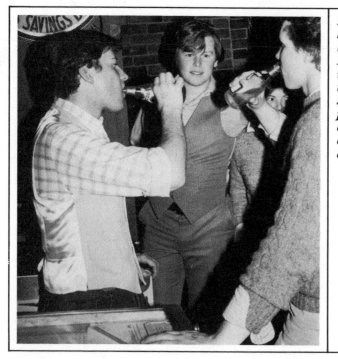

Boston teenagers hurriedly finish their beers before a Massachusetts law that raises the drinking age to 20 goes into effect. Such laws are being passed throughout the country in an effort to curtail drunk driving among teenagers.

UPI/BETTMANN NEWSPHOTOS

and a lower quality of goods and services. A common warning to new car buyers, "Never buy a car assembled on Monday," is based on the assumption that on Mondays workers perform poorly because they are feeling the aftereffects of a weekend of drinking. Similarly, some people believe that the inferior quality of many American-made products is related to the workers' use of alcohol and other drugs.

The enormous growth of alcoholism treatment facilities throughout the country testifies to the country's raised consciousness about the dangers of alcohol. Facilities now treating alcoholics include halfway houses, hospital-based detoxification units, free-standing outpatient (out of hospital) clinics, and private and public inpatient (in-hospital or free-standing) rehabilitation programs. As a result of the establishment of so many facilities, the field of alcoholism treatment has attracted professionals of many disciplines and led to the creation of the credentialed alcoholism counselor (CAC), an individual who specializes in counseling about alcoholism and ways of achieving and maintaining abstinence and sobriety.

An American bar in the late 19th century. In 1895 a survey of Chicago bars reported that on an average day the number of customers equaled half the city's population.

The self-help movement for alcoholics has also grown. Membership in Alcoholics Anonymous continues to rise, and new groups have been established, such as Drink Watchers, Women for Sobriety, and Pills Anonymous (a fellowship open to poly-drug abusers, i.e., those who abuse more than one drug).

At the same time, there have been national demands for improved health benefits for individuals suffering from alcoholism. The insurance industry has responded positively, and, as a result, more problem drinkers and alcoholics have sought professional help.

In addition to the expansion of and the increased accessibility to treatment methods, there have been stronger efforts by many groups to address the causes of problem drinking. There is also a new emphasis on *prevention*. These efforts deal with not only people on the verge of experiencing serious life problems, but also those individuals who, though not yet problem drinkers, have alcoholic tendencies. The goal is to reduce the number of people who drink to excess. The theory is that it is better to spend a little time and money now to fend off problems than to be faced with suffering,

A *Massachusetts policeman distributes a pamphlet describing new laws concerning drinking while intoxicated. Penalties for this crime can now include automatic suspension of one's driver's license and increases in automobile insurance premiums.*

AP/WIDE WORLD PHOTOS

illness, expensive treatments, and other burdens to the individual and society in the future.

Employee assistance programs (EAPs) have been established by companies for employees, who either voluntarily use available services or are referred to the programs by employers who have recognized that chemical abuse and/or other problems are interfering with workers' job performances. Thus far these programs have proven quite successful. Also, more and more school systems are running student assistance programs (SAPs), which are based on the same theory of early intervention to stop the formation of alcoholic and other self-destructive habits.

All of these efforts demonstrate a new commitment to and concern about alcohol abuse and alcoholism. Rather than blame "innate" character deficiencies and weaknesses, society has begun to accept the fact that people's behavior can be modified so that they are able to make more responsible decisions about personal use of alcohol.

Mothers Against Drunk Drivers was founded in 1980 by a woman named Candy Lightner after her daughter was killed by a drunk driver.

Actor Robert Walker being booked on a drunk charge in 1948. Three years later he was dead, the result of a lethal overdose of sedatives and alcohol. Using these depressants in combination is extremely risky.

CHAPTER 3

PROPERTIES AND EFFECTS OF ALCOHOL

Alcohol belongs to a family of chemicals that contains carbon (C) and hydrogen (H) atoms, and hydroxyl groups (OH) that consist of one oxygen (O) atom and one hydrogen atom. Ethyl alcohol, or ethanol (C_2H_5OH), is the form of alcohol found in alcoholic beverages. It is also used in sleeping pills and cold medications.

Methyl alcohol, or methanol (also known as wood alcohol), is a solvent, and ethylene glycol is used in antifreeze. Isopropyl alcohol is commonly used as rubbing alcohol and as an *antiseptic* (a substance that prevents the growth of microorganisms). All three are poisonous if drunk. Glycerin, another member of the alcohol family, is used in medications (such as nitroglycerin), paints, and cosmetics. Cholesterol (a fatty animal substance produced mainly by the liver) and Vitamin A are also types of alcohols. In general, the various other, non-ethyl, forms of alcohol are used in pharmaceuticals, detergents, perfumes, and other products.

In the manufacture of beverage alcohol products, especially distilled spirits, alcohols other than ethanol may be formed in very small quantities. Unwanted because of their undesirable effects on the body, these alcohols are considered impurities, and may be partly responsible for the syndrome of negative aftereffects of drinking known as the "hangover."

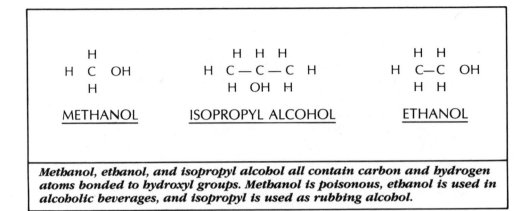

Methanol, ethanol, and isopropyl alcohol all contain carbon and hydrogen atoms bonded to hydroxyl groups. Methanol is poisonous, ethanol is used in alcoholic beverages, and isopropyl is used as rubbing alcohol.

Physical Effects of Alcohol

Alcohol affects the body in two ways. First, alcohol comes into direct contact with the linings of the mouth, esophagus, stomach, and intestine, where it acts as an irritant and an *anesthetic* (a substance that causes insensitivity to pain with or without loss of consciousness). The second set of effects occurs after alcohol is distributed throughout the body. Only 20% of the ingested alcohol is absorbed through the stomach; the other 80% is absorbed through the intestinal linings directly into the bloodstream. (Alcohol is not digested, or broken down, in these organs as foods are.) The alcohol molecule is very small, and thus reaches every cell in the body. However, because alcohol is water soluble (it dissolves in water), its concentration is greater in those organs that naturally contain more water.

When alcohol leaves the gastrointestinal tract (the stomach and intestines), it is carried first to the liver. There the alcohol is *metabolized*, or broken down, in various steps, each of which produces a unique molecule. This process continues until the original alcohol molecule is reduced to carbon dioxide (CO_2) and water (H_2O). On average, in one hour the liver can metabolize approximately .75 ounces of absolute alcohol, or the amount found in a 12-ounce can of beer, a 4-ounce glass of wine, or a 1.5-ounce serving, or "shot," of a distilled spirit such as whiskey.

While the alcohol is being metabolized by the liver, the heart continues to pump any unmetabolized alcohol throughout the body. The organ most sensitive to alcohol is the brain, which is responsible for coordinating and processing all thoughts, sensations, perceptions, and feelings. Alcohol's effects on these functions of the brain are the ones most noticeable to a person who has been drinking.

Any intake of alcohol produces an intoxication — the greater the amount ingested, the greater the effects. What alcohol does is *depress*, or slow down, the functioning of the body's cells and organs until they are less efficient. However, the effects of a drink vary according to how and under what conditions the alcohol is ingested. Most people can tolerate one drink per hour. This means that drinking at this rate produces little or no alteration in a person's ability to function. Therefore, sipping a drink over a period of one hour produces a minimal blood-alcohol concentration (BAC), or level of alcohol in the blood, and, in turn, yields a minimal intoxication. That same drink ingested in five minutes will produce a high peak BAC and, therefore, greater intoxication. Mood, emotional state, and body weight can also influence the effect of alcohol.

Therefore, in some ways, alcohol's effect on the brain is under the control of the drinker. Drinking slowly is preferable since it results in the greatest control. In addition, anything that slows down the body's absorption of alcohol, such as

Men drinking at a tavern. The effects of a drink vary according to how quickly and under what conditions alcohol is ingested. Most people can tolerate one drink per hour with no significant impairment of functions.

having food in the stomach, lessens the intoxicating effect. This is why it is recommended that people eat before or while they drink.

The first portion of the brain to be affected by alcohol is the cerebrum, or cerebral cortex. This is the outermost layer, which is responsible for coordinating functions of sensation, perception, speech, and judgment. Alcohol's effect on this portion of the brain produces slurring of speech and misjudgments in thinking.

The second portion of the brain affected is the cerebellum. Located in the back of the skull, the cerebellum is responsible for coordination and balance. Staggering, falling, and being unable to hold a lit match steadily are common examples of alcohol's effects on this area of the brain.

When the limbic system (the area of the brain beneath the cortex that is related to memory and sexual and emotional behavior) becomes affected, various exaggerations of emo-

ART RESOURCE

Most experts agree that moderate drinking under the appropriate circumstances can be beneficial to one's health. However, excessive drinking can distort perception and lead to irrational behavior.

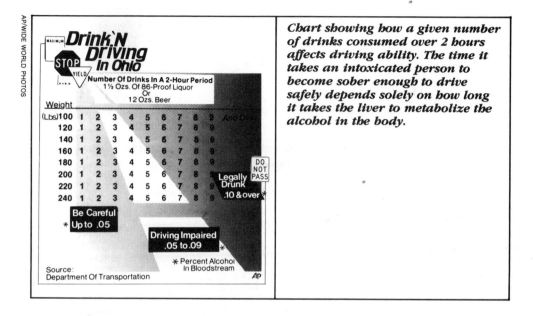

Chart showing how a given number of drinks consumed over 2 hours affects driving ability. The time it takes an intoxicated person to become sober enough to drive safely depends solely on how long it takes the liver to metabolize the alcohol in the body.

tional states may appear. They can range from boisterousness, verbal aggressiveness, and physically violent behavior in some persons, to quiet withdrawal and tearfulness in others.

Although alcohol depresses bodily functions, it often seems to act as a stimulant and to rid the user of his or her inhibitions. This is because the body responds to alcohol on several levels. Physiologically, alcohol always acts as a depressant, slowing down the speed at which nerves transmit information from one part of the body to another.

However, one of the functions of the cortex is to exert an inhibiting force on the emotions, which are organized in a lower portion of the brain. It is this control that gives people the ability to behave in a controlled or socially acceptable way. When the cortex is depressed by alcohol, not only are the senses and judgment affected, but the cortical controls on the midbrain (the area of the brain that connects the cerebrum to the cerebellum) are also depressed or relaxed. Therefore, during the period of time after the cortex is depressed but before the midbrain is depressed, the emotions are more easily expressed and the person feels exhilarated. As more alcohol is ingested, the limbic system areas are also depressed, and the alcohol drinker then feels the drug's depressant effects.

Acute Intoxication

The highly intoxicated individual will show severe signs of behavior dysfunction (abnormal, impaired, or inadequate behavior). As reflexes slow down, an inability to articulate words or even to stand may ensue. At this point alcohol has a growing effect on an even deeper or lower level of brain function — the arousal system. If enough alcohol is consumed, the drug user will fall asleep (a protective mechanism that prevents the individual from drinking even more); however, in extreme cases, he or she may lapse into a coma. For every individual there is a lethal dose of alcohol that is generally related to body size. Ingestion of a lethal dose causes death by inhibiting one or both of two basic life-sustaining reflexes that control heart rate and respiration and whose centers are located in the brain stem (the structure that connects the brain to the spinal cord). If enough alcohol is in the body, the depressant effects can permanently stop these centers from functioning and cause death. Therefore, "chug-

Destitute alcoholics such as these are an all-too-familiar part of the inner-city landscape. The damaging effects of chronic alcohol abuse are cumulative and have the potential to destroy almost every organ system in the human body.

AP/WIDE WORLD PHOTOS

a-lugging" a large amount of alcohol to the point of passing out, or mixing alcohol with one or more other depressant drugs can have obvious disastrous effects.

Because behavior is a function of brain activity, altering the brain by ingesting drugs also alters behavior. Intoxicated individuals are dangerous to themselves and to others. Because drinking increases confidence and diminishes abilities, the alcohol user is usually unable to judge accurately what he or she can or cannot do. Unfortunately, this misplaced confidence often leads people to judge themselves competent to carry out tasks that are beyond their abilities, as the many alcohol-related driving fatalities illustrate.

Chronic Abuse

It is not known how much and how often alcohol must be consumed to produce long-term physical effects in any given individual. It is clear, however, that as the quantity and frequency of consumption increase, the greater the risk that alcohol will begin to have cumulative effects on the functioning of the body.

Because alcohol is carried to every cell of the body, it has the potential to damage almost every organ system. The body has two limited defenses against this process. The first is to metabolize the alcohol more quickly, and therefore rapidly convert it to harmless substances. To some extent this action occurs in the liver. This amazing organ, which rids the body of many toxic substances, increases its efficiency in metabolizing alcohol by making greater use of a secondary system to supplement the primary alcohol-metabolizing system. There are, however, certain limits to how much the liver can do.

The body's second defense relies on a property common to all living organisms — the ability to adapt. There are two forms of adaptation — behavioral and physiological. With behavioral adaptation, an individual adjusts his or her behavior to compensate for the disruptive effects of excess alcohol in the blood. For example, a person who feels somewhat dizzy or unbalanced after having two drinks may intentionally or unconsciously lean against the wall for support. Someone who senses the effect that alcohol is having on his or her speech may purposely articulate words more carefully. Adaptations of this sort demonstrate an organism's attempt to

41

cope with the presence of a toxin it cannot escape, except by waiting for the alcohol to be completely metabolized.

Physiological adaptation takes place on a cellular level. A cell whose activity level has been slowed down by the depressant effects of alcohol will gradually speed up its operations until its alcohol-affected activity level returns to normal. This cellular adaptation is very important because it permits body processes to continue. Cells that adapt to the presence of alcohol seem to return to normal; their activity levels, when scientifically measured, are not very different from those of similar cells that have not been exposed to alcohol. In fact, the cells are not functioning in a completely normal way. To overcome alcohol's depressant effects and maintain their usual activity level, the cells must work harder.

Although alcohol's effects wear off as the drug is metabolized, the cells' operations do not slow down at the same rate. Therefore, in the absence of alcohol the cells' activity level undergoes a kind of *rebound effect*, characterized by a temporary above-normal activity rate. A rise in the activity level of any one cell cannot be perceived by the individual. However, when millions of cells rebound simultaneously, the effect is noticeable. In moderate drinkers the rebound may be felt as increased restlessness and/or anxiety. In heavy drinkers and alcoholics, the rebound may be so great that it produces severe shaking or tremors. Unable to stop or cope with the tremors, the heavy drinker often turns to alcohol, which, ingested in small to moderate doses, again depresses the cells just enough to stop the tremors.

In the short run, the body is able to adjust to alcohol's depressant effects. After a while, however, the body cannot maintain its equilibrium, and the more significant negative effects of alcohol manifest themselves. Many parts of the body respond to alcohol by becoming inflamed, or swollen. As a result of excess alcohol intake, inflammation (indicated in medical terminology by the suffix "-itis") commonly occurs in the heart (*myocarditis*), the stomach (*gastritis*), the liver (*hepatitis*), the pancreas (*pancreatitis*), and the nerves (*neuritis*). At first, these conditions may be reversible if alcohol intake is stopped. However, if alcohol intake continues, certain permanent, nonreversible changes may occur. These effects may include sterility, impotence, and *cirrhosis* (the loss of functioning liver cells and related dysfunctions).

Effects of Alcohol on the Fetus

When a pregnant woman ingests alcohol, the drug is distributed throughout her body. Thus, it crosses the placenta (the uterine lining that nourishes the fetus) and reaches the unborn child itself. Because the developing fetus is highly sensitive to toxic substances, alcohol can have very damaging effects.

Alcohol tends to slow down the movement of the fetus, just as it slows down the functioning of cells and organs. Studies of monkeys show that even small amounts of alcohol can interfere with the flow of oxygen to the fetus. Heavy exposure to alcohol results in a retardation of fetal development, as well as structural and functional disorders. The set of defects resulting from alcohol consumption during pregnancy is called *fetal alcohol syndrome* (FAS). Although not all of the symptoms are always found in a single infant, the range of disabilities includes incomplete hand development; facial deformities, including widely-spaced eyes, overgrown eyelids, thin upper lip, smaller distance between the upper lip and nose, and a flattened, or spread-out, nose; and abnormal brain development, leading to impairment of intellectual and motor (movement) abilities.

There is also evidence that pregnant women who consume smaller amounts of alcohol — as little as a drink or two

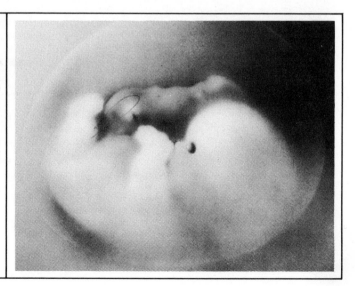

A seven-week-old fetus. Since even a moderate consumption of alcohol can interfere with the flow of oxygen to the fetus, sometimes resulting in serious birth defects, pregnant women are advised to abstain from drinking alcohol.

per day — may have more miscarriages and deliver smaller babies who grow more slowly and have more behavioral difficulties than babies of abstinent mothers.

The Myths and Realities of Sobering Up

On occasion, many drinkers ingest too much alcohol and become intoxicated. They then are usually concerned about how quickly they can be returned to sobriety. The desire to find a way to reverse rapidly the effects of alcohol has probably existed since alcohol was first drunk, and researchers continue to search for a solution to this dilemma.

Sobering up can only be achieved as the liver naturally detoxifies the body of alcohol. Taking cold showers and drinking black coffee, the two traditional approaches, do not work because they have no effect on the time it takes for ingested alcohol to be metabolized by the liver. The danger of these supposed remedies is that their stimulating qualities may *appear* to improve temporarily the alcohol abuser's condition. Overindulgers, awake but still impaired by the alcohol, may feel just alert enough to make the erroneous judgment that they are capable of normal functioning. They may feel able to drive an automobile, a conclusion that endangers both

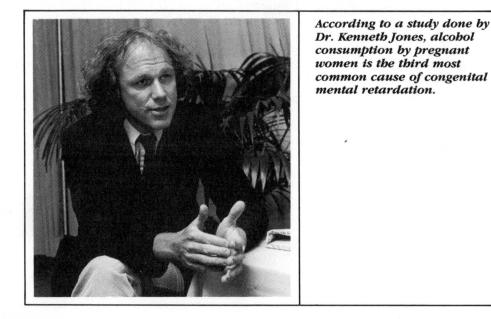

According to a study done by Dr. Kenneth Jones, alcohol consumption by pregnant women is the third most common cause of congenital mental retardation.

UPI/BETTMANN NEWSPHOTOS

them and the others around them. Today, it is still true that the only way to truly sober up and regain one's normal judgment, perceptions, and abilities is to rest or sleep until the effects wear off.

Alcohol and Other Drugs

Many people do not think of alcohol as a depressant drug, and therefore they are not aware of the fact that simultaneously using alcohol and other drugs can pose a serious health risk. When taken along with other depressant medications or illegal drugs, alcohol's depressant effect on the nervous system may be exaggerated. For example, taking alcohol in combination with an *antihistamine* (a drug used to treat colds and allergies and that also causes drowsiness), may severely depress bodily functions. In some cases combining drugs may cause an overdose, or even death. To make people aware of this danger, manufacturers are now required to indicate on the labels of medications that have depressant effects that these drugs should not be taken with alcohol.

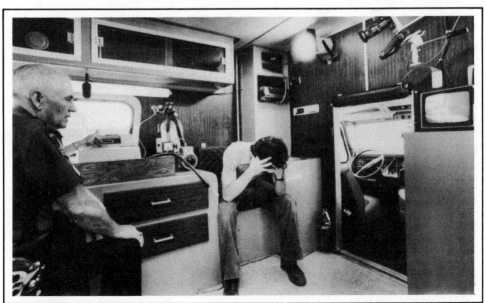

A Louisiana police officer questions a man suspected of driving while intoxicated. Substance abusers who mix alcohol with other depressants pose an especially dangerous threat on the highway.

Drinking patterns established in adolescence often carry over into adulthood. According to a recent national survey, an alarming percentage of high school students drink alcohol, often to excess.

CHAPTER 4

THE USE AND ABUSE OF ALCOHOL

Why do people drink, and when does their drinking become a problem? One way to begin analyzing these complex questions is to view them on a continuum. At one end is _social drinking_, here defined as the ritual or recreational consumption of alcoholic beverages at meals, parties, and special occasions. The alcohol may be an integral part of a ritual or serve to stimulate the appetite and help produce conversation, relaxation, and good feelings. Consumption is usually done at a slow pace, and involves small to moderate quantities, thus having very few negative consequences.

In the middle of the continuum is _alcohol misuse_, which applies to situations in which alcohol consumption causes obvious negative effects. The drinking individuals talk loudly and make inappropriate statements; sometimes they become aggressive or pass out. For people in this category, misuse generally occurs rarely or infrequently.

At the other extreme, _alcohol abuse_ is very similar to alcohol misuse, except that the excessive consumption occurs frequently or regularly.

There are no clear boundaries among these three categories; what looks like social drinking to one person may appear to be misuse to another person. To further complicate these distinctions, some people believe that even social drinking is as bad as alcohol misuse.

Socio-Cultural Influences

A number of factors contribute to an individual's decision to drink or not drink alcohol on any given occasion. The most obvious are those in the immediate environment, such as the nature of the occasion, one's personal desire and physical and mental state, and the degree of peer pressure. In addition to these are the *socio-cultural* influences, or all the historical and contemporary factors that contribute to an individual's attitudes, beliefs, and values regarding the use of alcohol. These may include religious, ethnic, and family customs.

There is evidence to suggest that the socio-cultural influences have less of an effect on an individual's decision to drink than do the circumstances of the immediate situation. In many cases, a person with a certain set of beliefs and expectations about drinking comes into contact with people who have different definitions of alcohol use. This individual may thus be put into a situation that demands social drinking at a level that he or she previously would have considered alcohol misuse.

Gifted actors Richard Burton and Elizabeth Taylor both struggled with acute drinking problems. Alcohol abuse contributed to Burton's death; Taylor entered a rehabilitation center and achieved sobriety.

Teenagers and Drinking

It is particularly important to understand what motivates adolescents to drink and misuse alcohol. This is partly because patterns acquired during this early period frequently carry over into adulthood. The factors that trigger members of this group to misuse and abuse alcohol, however, are especially complex.

A unique period of growth and development, adolescence marks the transition from childhood to adulthood. It is characterized by increased mobility and freedom, demands for more independence, and increased capabilities in many areas. Adults begin to expect an adolescent to be more responsible. During this period, adolescents strive to become adults and thus begin to imitate adult behavior, which often includes drinking. Studies show that as young people in the United States progress through adolescence, increasingly greater numbers of them consume alcohol. A recent national survey found that by the 12th grade, 93% of boys and 87% of girls have tried alcoholic beverages. Researchers have concluded that there are at least five major reasons why adolescents drink.

A highway "safety stop" serves coffee to weary travelers. Although caffeine can combat drowsiness and enhance awareness, it does not help an intoxicated person drive safely.

Adult/Parental Modeling. Because adolescents aspire to adulthood, they often adopt the drinking styles and patterns of their parents and other important adults in their lives, including societal heroes. Many parents expect their children to learn how to drink and to "handle" alcohol in an adult manner.

Curiosity and Experimentation. Although when they were younger they may have sampled alcohol at home, during family affairs, at special events, and on religious occasions, many adolescents do not know how to drink or what alcoholic beverages really taste like. Out of curiosity, they may taste alcohol in its various forms, in different combinations, and often in increasing quantities. When overindulgence occurs, it often represents the novice's attempt to define his or her capacity, as well as to experience how it feels to be intoxicated. Unlike adult abusers, who deliberately become intoxicated, inexperienced drinkers usually do not intend to get drunk but may become so when experimenting with alcohol.

Peer Pressure. Every group, and especially groups of adolescents, is susceptible to peer pressure. Such pressure can be real or imagined. An example of real peer pressure is when an adolescent is told, "Either drink with us or find yourself

After overcoming her addiction to both alcohol and tranquilizers, former First Lady Betty Ford opened the Betty Ford Center for Chemical Dependency in 1982 to help others achieve sobriety.

AP/WIDE WORLD PHOTOS

another group to hang out with." Imaginary peer pressure occurs when adolescents generate their own pressure as a result of their fear, often unfounded, of being rejected by the group. Unfortunately, both pressures may ultimately lead to unwise drinking habits.

Enjoyment. The belief that a social activity cannot be enjoyable unless it includes the consumption of alcohol is not uncommon in our society. Many people think that they will not be able to enjoy a party, a movie, or a football game unless they are feeling the effects of alcohol. Teenagers can be especially susceptible to this point of view because of the peer pressure discussed above. When an adolescent hears his or her peers say, "It is more fun with beer," or "The party will be boring if they aren't serving booze," he or she is apt to conclude that alcohol is an integral part of these social occasions.

Emotional Turmoil. An immediate, short-term effect of alcohol is mild *tranquilization*, or a euphoric sense of well-being. Feelings of uneasiness, tension, and fear, as well as more severe symptoms of panic and depression, are reduced. This is why many people, including teenagers, find alcohol use beneficial. During adolescence, many teenagers experience considerable personal turmoil as their roles, emotions,

Dr. Glenn Weimer, a Texas psychologist, established a crisis hotline in response to the rapidly rising teenage suicide rate. There is a direct correlation between alcohol abuse and suicidal tendencies in teenagers as well as in adults.

expectations, capabilities, and desires change. It is understandable that, for these individuals, using alcohol to achieve its tranquilizing effects is a great temptation. Drinking alcohol, however, merely desensitizes teenagers to their problems for a short time. Drinking presents no genuine or long-term solutions, and only leads to more and more serious difficulties.

Alcoholism

Alcoholism is the term used to describe the most severe misuse of alcohol. It is a condition that carries powerful negative connotations and social stigmas. Although alcoholism differs from individual to individual, all alcoholics have two things in common: they drink alcoholic beverages, and their alcohol-influenced behavior gets them into trouble.

Alcoholism is a *disorder*, or a pathological (disease-based) condition, of both behavior and physiology. The alcoholic's alcohol-seeking behavior is abnormal, and his or her body reacts to it poorly. Like many disorders, it is one that is caused by a number of biological, psychological, and social factors.

An engraving showing a man undergoing withdrawal from alcohol in the dungeon of a 19th-century prison vividly captures the primitive methods once used in the treatment of addiction. In cases of acute alcoholism, this kind of abrupt withdrawal can be fatal.

THE BETTMANN ARCHIVE

Biological Factors. The observation that alcoholism tends to run in families does not prove that it is inherited. Family members may learn to drink simply through observation and imitative behavior. However, a number of recent research findings suggest that there are strong genetic influences that can lead to alcoholism. Several studies have been able to separate the genetic and environmental influences that affect this disorder. For example, there have been studies of identical twins (who have exactly the same genetic material) reared apart and of children adopted and raised by foster parents.

The best known study was conducted in the 1970s by a research team that examined the records of two groups of adopted Danish children. The first group consisted of 55 children whose biological (natural) parents were alcoholics; the second group was made up of 78 children whose biological parents were not alcoholics. Some of the children from each group were raised in homes where at least one of the foster parents was alcoholic, and the rest lived with foster parents who were not alcoholics.

Of the first group (children of alcoholics), 10, or 18%, became alcoholics, and of the second group (children of nonalcoholics), only 4, or 5%, became alcoholics. This was true despite the fact that only 7% of the foster homes of children of alcoholics had an alcoholic foster parent, though 12% of the foster homes of children of nonalcoholics had an alcoholic foster parent. The researchers were able to conclude that children of alcoholic parents, even after being separated from their parents as early as the first six weeks of life, were 3.5 times more likely to become alcoholic than children of nonalcoholics.

But what does a genetic theory of alcoholism imply? The most extreme view is that if alcoholism is genetic, it is beyond all control of people as to whether or not they become alcoholics. A more moderate and far more accepted theory is that of *genetic predisposition*, which holds that alcoholism can be triggered by certain inherited conditions. There are a number of explanations for what causes the predisposition. Some researchers believe it may be a physiological or biochemical imbalance that is corrected by consuming alcohol. Others think that alcohol actually takes the place of a chemical deficiency in the alcoholic's body. And still other researchers explain the predisposition as being a progressive inability of the alcoholic's body to metabolize alcohol normally. Whatever the cause of the susceptibility, the predisposition theory leaves room for some notion of personal responsibility, because the person must decide to drink in order for the triggering effect to occur.

Disease. Related to the theory of genetic predisposition is the view of alcoholism as a disease. First characterized as a disease in the 19th century, alcoholism was officially labeled as such in 1956 by the American Medical Association, and again in 1969 by the American College of Physicians. In 1960 E. B. Jellinek, a pioneer researcher and theorist on the subject of alcohol-related disorders, described several types of alcoholism, two of which he characterized as diseases: *gamma* alcoholism and *delta* alcoholism. Common to both types is tissue *tolerance*, whereby the body becomes less susceptible to the effects of alcohol as usage continues, and *metabolic adaptation*, whereby the body is able to alter its metabolic processes (all the processes that are involved in the cells' transformation of energy and matter to sustain life) to adapt to the presence of alcohol. Both tolerance and metabolic adaptation lead to physical dependence and the characteristic withdrawal symptoms that occur when alcohol intake ceases. (Physical dependence involves habitual craving for a drug.)

The *gamma* alcoholic, who exhibits an inability to control the amount of alcohol ingested once drinking begins, alternates between uncontrolled bouts of drinking and periods of abstinence. In contrast, *delta* alcoholics are rarely able to abstain from drinking. They generally drink smaller quantities of alcohol than *gamma* alcoholics and are therefore better able to function, but their drinking frequently continues throughout the day.

A more recent and popular conception of alcoholism is to see *all* alcoholics as victims of disease. According to this viewpoint, alcoholism is progressive and leads to serious complications if there is no intervention. Therefore, even those alcoholics who do not initially exhibit the symptoms of disease will suffer from them in the future. After interviewing over 1,000 male alcoholics, Jellinek described a likely progression of changes in thought, behavior, and drinking that occurs in people who move into an alcoholic drinking pattern. As the need for relief from psychological, and sometimes physical, pain and discomfort increases, the individual drinks more and more frequently, rationalizing the increased intake in various ways. Each new level of drinking and its consequences may produce a new series of difficulties relating to the drinker's health, work, friends, and family.

Although the progressive-disease view of alcoholism is widely accepted, it is not without flaws. It does not account for those alcoholics who do not progress through the usual stages or do not develop the disease characteristics described by Jellinek. In addition, the theory does not explain why some alcoholics are able to reduce their alcohol intake to a level that does not produce life-disrupting complications. It also fails to explain why some alcoholics are able to give up drinking spontaneously and completely.

Psychological Factors. For many years it was fashionable to explain alcoholism by asserting that alcoholics suffered from personality disorders. Recently, however, experts have

UPI/BETTMANN NEWSPHOTOS

Kansas City Royals catcher and recovering alcoholic Darrell Porter holds a bottle of non-alcoholic sparkling grape juice while celebrating a team victory. Porter missed part of the 1980 season while he was treated in an alcohol rehabilitation center.

come to agree that there is no such thing as a distinct "alcoholic personality." Evidence points to the fact that prior to becoming addicted, genuine alcoholics have some shared personality disorders and neuroses that can include narcissism (self-love or self-admiration), paranoia, and obsessive-compulsiveness. Alcohol may have a unique effect on these individuals.

Although these pre-existing personality types may influence alcoholic behavior, they do not necessarily cause alcoholism. (There are many individuals with similar psychological problems who do not become alcoholics.) However, as an individual recovers from alcoholism, the problems that existed prior to addiction will not only influence the type of recovery but will also strongly determine how the person readjusts to society.

Learning. Some researchers claim that drinking alcohol is a learned behavior that is governed by the normal principles of learning. Uncontrolled drinking is viewed as a destructive habit that is exhibited under negative conditions, such as during times of great stress. The behavior is reinforced by the positive consequences that immediately result from the alcohol — relief from anxiety and feelings of increased ability and power. Although drinking has both positive and negative effects, the negative ones usually occur after the positive ones and have less influence on the learning pattern.

This theory of learned behavior has generated controversy about how to treat the alcoholic. Learning theory predicts that an acquired, or learned, behavior can be unlearned, or at least modified. Therefore, alcoholics should be able to learn to control their alcohol intake and not need to stop drinking completely. This position does not take into account the possibility that the addiction to alcohol causes certain, though unidentified, physiological mechanisms to come into play.

In fact, evidence indicates that only 5% to 10% of alcoholics are able to return to nonproblem drinking for substantial periods of time. The rest of them quickly relapse into uncontrolled drinking. It is commonly agreed that abstinence is the only practical approach that will help alcoholics return to productive lives.

The Holistic Theory. This theory states that alcoholism is a way of life in which interactions with people, places, and events are influenced by the potential use and actual use of alcohol. In some ways this theory is a more sophisticated version of the learning theory, which tends to claim that alcoholism is merely a bad habit. According to the holistic theory, the alcoholic's needs, perceptions, judgments, expectations, and anticipations of self and others are all colored by the consequences of drinking. Therefore, it is not alcoholism alone that is a disease but also all the external and internal pressures — environmental, psychological, and biological — that bear on the alcoholic. This in part explains why abstinence alone does not help many alcoholics become productive individuals. The holistic approach recognizes the importance of treating the psychological, sociological, nutritional, and physical consequences of excessive drinking. Therefore, in addition to abstinence, treatment includes diet, exercise, the improvement of social skills, and the establishment of a total social support system.

A crowd at a New York nightclub. Some experts believe that alcoholics can learn controlled drinking. Evidence indicates, however, that only 5–10% of alcoholics are able to learn how to control their drinking.

EVERY BREATH YOU TAKE, EVERY MOVE YOU MAKE, WE'LL BE WATCHING YOU.

THE POLICE

METROPOLITAN POLICE

This holiday season, the Metropolitan Police will be administering breathalyzer tests at road blocks all over town. And if you don't pass, you just might pass the next two years in jail. So enjoy the holidays, but if you're drinking, don't drive. Unless you want to pick up a record you really can't enjoy.

Michael S. Dukakis
Governor
William J. Geary
Commissioner

This poster, warning teenagers against driving while drunk, is one of many prevention efforts aimed at minimizing the possibility that any problems associated with drinking will ever occur.

CHAPTER 5

PREVENTION, INTERVENTION, AND TREATMENT

*T*oday's medical community and mental health care professionals address the problems that result from alcohol abuse and alcoholism in three major ways. Primary *prevention* efforts are intended for individuals who drink alcoholic beverages but do not have alcohol-related problems, and for those people who do not yet drink alcohol. The goal of prevention is to minimize the possibility that problems with drinking will ever occur. Secondary prevention, or *intervention*, focuses on alcohol users whose drinking is causing problems and attempts to keep these problems from becoming severe. Tertiary prevention, or *treatment*, deals with drinkers and alcoholics who are experiencing severe problems as a result of their drinking. The goal of treatment is to help these individuals stop drinking and return to a productive life.

All three forms of prevention include an educational component that is designed to impart information regarding the nature of alcohol, the consequences of using it, and alternatives to drinking, regardless of what motivates it.

Prevention

Why should society make an effort to prevent alcohol abuse among people who are not experiencing any alcohol-related difficulties? Studies suggest that the annual increase in new alcoholics is greater than the number who are successfully treated, and that it is from the population of nonproblem drinkers that new problem drinkers emerge.

Early approaches to prevention relied heavily on scare tactics, especially in efforts directed at young people. More recent approaches, however, have focused on clarifying values and on changing attitudes. Furthermore, the latest preventive strategy is to capitalize on America's newfound awareness of the importance of proper nutrition and exercise.

Intervention

The main goal of intervention is to make problem drinkers aware of the connection between their drinking and their physical health, personal relationships, job performance, and economic circumstances. Many problem drinkers attempt to deny these connections; if they have not, in fact, "hit bottom"

Alcoholism counselors conduct group therapy sessions as part of the program at the De Paul Rehabilitation Hospital. This kind of treatment is often essential for alcoholics attempting to return to sobriety.

yet, they may not be desperate enough to seek help on their own. In light of this resistance, intervention must be tailored to the individual and his or her circumstances.

Also known as *confrontation*, intervention treatment presents the problem drinker with clear evidence of his or her self-destructive behavior. The confrontation might be produced by a spouse who threatens divorce, by an employer who threatens a loss of job, or by a physician who points out the profound medical consequences of continued drinking. Such confrontations are frequently successful in moving the problem drinker to action. In general, the earlier that intervention takes place in a drinking career, the more successful it is.

Intervention programs are discussed in more detail later in this chapter.

Treatment

In most cases, treatment for alcoholism begins with the stopping of drinking, preferably under medical supervision. This

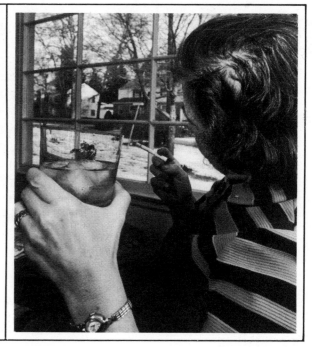

Traditionally, society has judged heavy drinking among women extremely harshly. This has led many women alcoholics to drink in the privacy of their homes, often alone. It has also made many women particularly reluctant to admit their problem and get professional help.

process, known as *detoxification*, usually takes about five days. A person may then transfer to a longer-term (four weeks to three months) residential rehabilitation facility or to an outpatient clinic. Sometimes homeless persons can move to an alcoholism halfway house, where they live for periods ranging from six months to about two years.

Most alcoholics are treated for both their medical and psychological problems. Psychological treatment often includes individual and group counseling, alcohol-education classes, and AA meetings. When appropriate, family members participate to help clarify problems and build a supportive environment.

Alcoholism treatment programs are discussed further in chapter 6.

Working with Special Populations

There are many different types of people with drinking problems. Because many of these people belong to groups that are apt to face other problems as well, it is important that these problems be dealt with during treatment. In many cases treatment must be adapted to special socio-cultural circumstances. Also, some groups have traditionally had little access to treatment. Interestingly, it is the demand for improved access that has sometimes prompted groups to acknowledge that drinking can be a problem for their members, not just for "others." Over the past 15 years, treatment practices have come to address the needs of a number of special groups.

Women. Alcoholism has traditionally, but incorrectly, been viewed as a male problem. Although social drinking among women has gradually become acceptable, excessive drinking has not. Most women who defy convention and drink excessively are criticized more severely than men with similar drinking problems. Consequently, women who drink heavily tend to do so alone and at home. Also, they often feel as though they have no one to turn to for help.

In the past a great majority of alcoholism treatment programs were geared to men, and thus most staffs were ill equipped to deal with women's special needs. Frequently, women had different social roles and patterns of emotional responses than men, and they most often were responsible for raising the children in a family. Only recently have programs started to acknowledge these differences and care for alcohol-abusing women in a more sensitive and appropriate manner.

Ethnic Groups. It is comparatively easy to recognize that Russian, Vietnamese, Korean, or other recent immigrants with alcohol problems need more than just alcoholism counseling to recover. The stress and difficulties related to adopting a new country and language demand special attention. However, it is not always acknowledged that well-established minority groups in the United States also need special help.

Native Americans, for example, are apt to feel displaced, disenfranchised, and rejected in their own homeland. Their cultures often emphasize beliefs and behaviors not understood or accepted by mainstream America. In recognition of

Native Americans gather at the Indian Lodge, an alcohol treatment facility in Los Angeles. Special alcoholism programs for this and other ethnic groups have been established throughout the United States.

these differences, special alcoholism programs for Native Americans have been established throughout the United States. These programs often incorporate traditional *shamans*, or medicine men, and American Indian rituals into more conventional alcoholism counseling.

Similarly, many Hispanic alcoholics need contact with people who understand their own language and respect their religious beliefs and cultural habits. Many Hispanics respond better to *espiritistas*, or spiritual healers, than to mainstream professionals.

Many blacks live at the poverty level and experience economic and social discrimination. Treatment staff must understand how these factors contribute to attitudes about alcohol. They must also consider that most ethnic groups comprise various sub-categories. Native Americans are members of many different tribes; Hispanics can be Puerto Rican, Cuban, or Chicano; blacks can be West Indian, African, or

Comics such as this one are part of prevention efforts aimed at making young people aware that they are not immune to the risks of alcoholism. Tragically, many teenagers in the United States suffer from serious drinking problems.

AL-ANON

Afro-American. Each of these subgroups may view the world in a somewhat different way, and these differences may have serious implications for treatment.

Adolescents. All people go through a prolonged transition period between childhood and adulthood that is often characterized by physiological and psychological turmoil. Because of this, both the medical and mental health fields often require professionals to undergo special training before they can treat adolescents. Most alcoholism treatment centers, however, are designed to handle the adult population and do little to attract and keep these younger clients. Teenagers with alcohol abuse problems find it difficult to relate to older staff members, especially those who have a rigid philosophy of abstinence. Therefore, adolescents often resist treatment that is really more suited to confirmed alcoholics.

Until more alcohol treatment centers develop appropriate approaches, many of these young alcohol abusers — and even those teenagers who abuse other substances — may find that they are better served by youth-oriented drug treatment centers and school-based prevention and intervention programs, such as the student assistance programs. These programs, modeled after the employee assistance programs found in industry, are usually staffed by young social workers and have a style and philosophy that attracts many teenagers.

Young Children. Special attention is required for young children who are members of families with alcohol problems or who are themselves alcohol abusers. Treatment is best handled by specialists in child psychology or psychiatry. Although knowledge of alcoholism and the alcoholic family is necessary, perhaps the most crucial factor in this kind of treatment is the basic ability to understand and relate to children.

The Elderly. Old people who are alcoholics or alcohol abusers fall into two categories: those with drinking problems that began before they became elderly, and those whose drinking problems are situational reactions to advanced age and its problems, such as isolation, bereavement, ill health, lowered self-esteem, and the side effects of medication. The latter group is by far the larger one. Nevertheless, the elderly person with a situational drinking problem is likely to be treated in traditional fashion by a mainstream alcoholism agency. Although many mainstream clinics do successfully care for elderly people with drinking problems, old people with situational drinking problems are best treated in specialized programs whose staff is trained to work with this group. When other problems that afflict the elderly are dealt with through counseling and increased social involvement with peers — a kind of treatment more frequently found in special programs for the elderly — the alcohol abuse usually subsides.

The Multi-Disabled. This group includes alcoholics or alcohol abusers who have one or more constitutional disabilities, such as visual or hearing impairments or the impaired use or loss of a limb. Like the elderly, these people also fall into two categories: those whose drinking problems preceded their disability, and those whose drinking problems developed as a response to it. Although many special programs to help the disabled already exist, there has been little financing of programs or agencies to address their alcohol problems.

This senior citizen has just completed a bike ride from New York City to San Francisco. The obvious physical vigor the man exhibits contrasts dramatically with the conditions of some of his peers, whose loneliness and ill health often lead to drinking problems.

Today, many alcoholism clinics are accessible by wheelchair. When disabled people reach a clinic, however, they often find that few staff members understand and/or are motivated enough to deal with the complications brought about by the handicaps. It would seem that the logical solution would be for agencies already helping the disabled to expand their services to include alcoholism and alcohol abuse treatment. Unfortunately, though members of this group do frequently suffer from drinking problems, thus far most agencies for the disabled have not taken on this task.

Homosexuals. As American society has become more open and liberal towards homosexuality, the gay population has become more outspoken in its efforts to have special services made available to its members. Alcohol abuse and alcoholism are not infrequent problems for homosexuals, and some people believe that drinking problems cannot be treated adequately if the homosexuality of the individual is ignored. Some counseling and treatment centers for gays are able to address simultaneously both the individual's sexual identity and his or her alcohol problem. For some homosexuals with drinking problems, sexual orientation is not a critical issue and they may be helped by traditional alcoholism agencies.

In some cases alcohol abuse may actually result from a person's attempt to deal with his or her homosexuality. For such individuals, drinking may thus be a secondary problem. If he or she is to benefit from alcoholism counseling with a non-homosexual counselor, the counselor will have to be a person who is comfortable with homosexual clients.

Multiple-Substance Abusers. Many alcoholism treatment centers have adapted to the reality that most people with alcohol problems also abuse other substances. Since the basic philosophy of alcoholism treatment — to encourage abstinence — can be applied to treatment for all substance abuses, many counselors are comfortable working with people who abuse other substances. Although this task is relatively easy when the other abused substances are in the sedative-drug family, whose effects are similar to alcohol's effects, working with stimulant abusers can be more difficult, because the stimulant produces effects that are quite different from alcohol's effects.

Most alcoholism facilities do not admit clients who are taking *methadone* (a synthetic opiate that produces effects similar to morphine's) as a way of ending heroin addiction. The basic philosophy of methadone maintenance, which supports the continued use of one drug to help end an addiction to another drug, conflicts with the philosophy of abstinence practiced at alcoholism treatment centers. Although most heroin addicts freely abuse alcohol, they are usually better cared for by drug-abuse agencies that have access to ambulatory detoxification procedures, therapeutic (long-term residential) communities, and methadone maintenance.

Cocaine abusers, who are rapidly increasing in number, often drink alcohol because of its depressant effects, which temper cocaine's stimulating effects. Apparently, most multiple abusers seek help for their primary drug problem, rather than for their drinking.

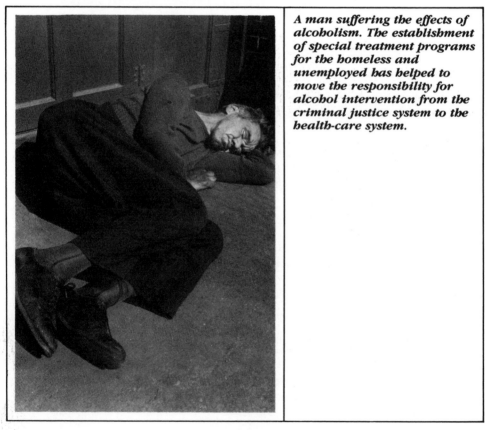

A man suffering the effects of alcoholism. The establishment of special treatment programs for the homeless and unemployed has helped to move the responsibility for alcohol intervention from the criminal justice system to the health-care system.

THE BETTMANN ARCHIVE

The Indigent. This group includes the homeless, unemployed, disaffiliated people who live in doorways, alleys, and basements; the occupants of single-room-occupancy hotels; and the deinstitutionalized mental patients who are barely able to manage daily life. Seldom are they enthusiastically welcomed by traditional treatment centers. However, the establishment of special programs for the indigent has helped to move alcohol intervention from the criminal justice to the health-care system. Some people in this group, no longer arrested for public intoxication, have been treated in social rehabilitation centers and have become productive individuals. Some have even become alcoholism counselors and have returned to their old neighborhoods to try to help others.

Ex-Offenders. People who have been in jail, some of whom may still be on parole, are not readily accepted into the mainstream of American life. To build their self-esteem, many band together in ex-offender groups, where they often become aware of the need to address alcohol abuse and alcoholism problems. As a result, many of these groups have developed successful counseling services that treat the problems ex-offenders have with alcohol and other substances of abuse.

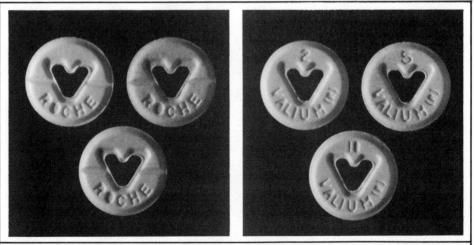

Many alcoholics also develop dependencies on drugs such as Valium. In order to treat multiple-substance abusers successfully, a rehabilitation program must address the special problems of each addiction.

Traditional treatment programs for alcohol problems suit a wide range of people. It is clear, however, that they do not and often cannot serve the majority of people with special problems. It is not only program philosophies that are the obstacles to treatment. It is also the attitudes of program staff, a lack of information on the needs and characteristics of the members of these social subgroups, and a lack of training in how to meet their special needs.

Besides bringing together people with alcohol problems, special programs allow those who share other major problems to meet and confide in each other. Many members of these minority groups are consoled by knowing that they are not the only ones experiencing these difficulties. Indeed, special programs for special populations have successfully reached and treated many people who otherwise might not have found a place for themselves in the mainstream alcoholism treatment programs.

A recovering alcoholic chats with his clientele at the non-alcoholic "dry bar" he opened in Illinois. His establishment caters to people from all walks of life who want an environment in which they can relax and socialize without being pressured or tempted to drink alcohol.

Education

In dealing with alcohol-related problems, one group whose needs should be given special consideration is young people, many of whom drink only occasionally or not at all. At some point many of these young people will make the decision to drink. Is there anything society can do to influence this group's decision-making process to minimize future alcohol-related problems?

There is now a consensus that something can be done. Properly run prevention and educational programs can guide young people through the series of decisions they will have to make about drinking. During the 1970s the National Institute of Alcohol Abuse and Alcoholism supported hundreds of pilot prevention programs around the country. Several outstanding programs were developed and selected for use as models.

What all these models have in common is the assumption that people are generally uninformed about alcohol and its short- and long-term effects. In addition, the models are based on the premise that even when individuals might wish to avoid using alcohol, they lack the personal strength and social support to act on their convictions. Therefore, the aim of these prevention programs is to inform people about alcohol and to create an environment that supports positive decision making about its use.

Certain information should be conveyed to current or future social drinkers. People should know that alcohol, though legal to possess and use, is a potentially dangerous drug. They should also know about the relationship between the amount of alcohol consumed and its effects on the body, especially the factors that affect the brain and behavior. To reinforce these points, programs should explain alcohol's effects on sensation, perception, judgment, and reaction time. They should stress the dangers of overdose and of driving while intoxicated.

Legislation

In addition to the spread of alcohol education programs, society's growing concern about the negative consequences of drinking has been reflected in changes in laws dealing with the drinking age, taxes, product packaging, advertising, and the drinkers' responsibilities for their actions.

Drinking-Age Laws. These laws were originally designed to protect children from the effects of alcohol, though today they are also seen as a way of protecting society from the consequences of alcohol abuse — especialy drunk driving. In 1985 the federal government began urging states to enact laws that would raise the drinking age to 21, and since then many have done so. Some people have suggested, however, that making alcohol an illegal drug for more young people only encourages the use of other, perhaps more dangerous, drugs.

Taxes. Purchasers of alcohol pay substantial taxes. Some modern-day prohibitionists believe that access to liquor can be reduced by increasing its cost, and thus support tax increases on alcoholic beverages. Others, however, contend that higher taxes will merely reduce purchases of social drink-

A 19th-century woodcut satirizes a law that tried to hold bar owners responsible for the drunken, sometimes criminal actions of their patrons. During the 1980s, many states have enacted similar laws, placing responsibility for the consequences of over-indulgence on the server in addition to the drinker.

ers, but will not really affect the purchase patterns of heavy drinkers, problem drinkers, and alcoholics.

Warning Labels and Lists of Ingredients. Many people believe that users of alcoholic beverages should be warned of the dangers of inappropriate alcohol use and be provided with a list of ingredients. Consumers would thus be able to make purchases based on an evaluation of each beverage's compatability with his or her particular nutritional and health needs. The assumption is that with this additional information, more people will decide to reduce or eliminate their consumption of alcohol.

Advertising. There is increasing pressure on the United States Congress to pass legislation forbidding the advertising of beer and wine on television. (Presently, networks voluntarily withhold liquor commercials.) The alcoholic beverage industry defends its advertising by saying that it is designed only to influence people who already drink. Proponents of the proposed legislation contend that the advertising not only contributes to increased consumption, but also portrays drinking as a necessary part of having a good time, feeling accomplished, and/or being accepted by others. Looking beyond the advertised product and examining the context and the way in which the product is presented, legislation supporters believe that through the commercial a powerful — and incorrect — message is presented — that the use of the product can enhance such highly valued characteristics as sexiness and self-esteem.

Responsibility. Another trend related to prevention is taking place in the courts, which with increasing frequency hold drinkers responsible for the consequences of acts performed under the influence of alcohol. Neither alcoholism, intoxication, nor amnesia is so readily recognized as the excuse it once was.

The responsibility for behavior due to intoxication has been spread even more widely. In many states, tavern owners may be held responsible for the acts of customers who leave their establishments in an intoxicated condition. Indeed, in New Jersey, hosts and hostesses can be held responsible for the actions of an intoxicated guest even after the person has left the household.

Intervention Programs

Intervention, or secondary prevention, is directed toward those drinkers whose alcohol use has created special problems. It is intended to prevent existing problems from worsening. Traditionally, intervention was used to help people whose problems were already serious, though recently these techniques have been applied more broadly to social drinkers who may experience only occasional difficulties. For example, individuals convicted of driving while intoxicated or driving under the influence may be required to attend special classes that educate them about alcohol, its abuse, and its effects on driving. In many cases, participation in these classes includes undergoing a clinical assessment for alcohol abuse and alcoholism.

Recovering alcoholics drinking coffee in the recreation room of a rehabilitation center. The longer it takes a person with a drinking problem to seek help, the more difficult his or her recovery will be.

Current knowledge holds that the earlier the intervention in a potentially troublesome drinking career, the better the chances of reversing the trend and overcoming the problems. These intervention programs do not necessarily begin by focusing on alcohol or other substance abuse problems. Instead, they first concentrate on a person's failing or deteriorating performance, regardless of the cause. In some of these programs, however, it has been found that in as many as 60% of the cases, alcohol or other drugs contribute to the performance problems that bring the people into the program.

Intervention programs have been introduced into industry, where they are called employee assistance programs (EAPs) as well as into schools, where they are called student assistance programs (SAPs). In both EAPs and SAPs, supervisors and teachers are alert to deteriorating performance and behavior change. When this is detected, students or employees are referred to a counselor for assessment. Studies indicate that these interventions have had some success.

It should be noted that interventions need not be carried out only by teachers, employers, or professional counselors. A remark by a friend or family member is an intervention because it serves to alert the person to the knowledge and concern that other people have about his or her drinking behavior. Any single intervention need not be successful to have an impact; it may take several interventions before a person begins to take action. This is because, when confronted with an aspect of their behavior they do not wish to recognize, people often deny that it exists at all. This initial response is often a reflection of an unconscious attempt to minimize a problem's significance and protect the ego against criticism. With sensitive, caring, and supportive intervention, the problem drinker will eventually feel comfortable enough to admit that he or she is having alcohol-related problems.

One final important point about early intervention is that it need not be overly forceful. For example, it may be more helpful and tactful to approach an alcohol-misusing peer by saying, "You don't seem to be your usual self lately. Is everything okay? Can I be of help, even if it is just to listen?" A somewhat more direct comment that still does not make the person feel cornered might be, "I notice you have been drinking quite a bit lately, and more often, too. Is this getting to be a habit or are you under some greater stress these days?"

Former alcoholic Phil McTigue stands by the service station he established as part of an alcoholic rehabilitation program. The station, which is run by recovering alcoholics, is just one example of how rehabilitation programs can aid recovery.

CHAPTER 6

TREATMENT FOR ALCOHOLICS

It is now generally acknowledged that in the United States there are approximately 10–13 million people in various stages of alcoholism, many of whom have never been in treatment for alcoholism and, therefore, are undiagnosed. A majority of these people are purported to be employed. The cost to society of such widespread alcoholism is estimated to be as high as $120 billion annually. This figure includes not only the actual costs of treatment paid by insurance companies, but also the losses from work-related lateness, absenteeism, excessive health benefits, lowered productivity, and accidents, as well as those losses resulting from unemployment and disability insurance payouts.

Alcoholism encompasses a lifestyle characterized by an inability to control intake and, in later stages, by periodic relapses following attempts to remain abstinent. Repeated cycles of abstinence or reduced drinking alternating with drinking bouts that lead to negative consequences occur for two reasons: because problem drinkers minimize or deny the severity of the situation; and because they almost desperately hold onto the belief that they can manage the problem themselves.

No segment of society is immune from alcoholism. This disorder has no respect for age, sex, race, religion, or national origin. Current evidence even suggests that for those cultural groups with supposedly low rates, such as Italians and Jews, alcoholism rates are higher than expected.

Goals of Treatment

In general, arriving at a definition of a treatment goal has depended on how the problem of alcoholism itself is defined. If, as many people think, the alcoholic is the victim of a disease, the goal is total abstinence. In other words, if the body cannot properly metabolize alcohol, or if there is a biological or biochemical defect, then the body lacks the mechanisms that permit any drinking.

Some people, however, define the alcoholic as a person who has developed a strong habit triggered by external and internal stimuli. To them, an alternative to abstinence is de-conditioning, or unlearning, the habitual response. According to this belief, treatment goals include learning how to drink moderately.

In most cases, abstinence does seem to be the best alternative. However, this goal is absolute in nature. If someone who is trying to control an alcohol problem slips and has even one drink, he or she is likely to feel a sense of failure that will undermine the resolve necessary to remain sober. Although the moderate-drinking approach does not work as well as supporters believe it should, it is flexible and therefore avoids this one drawback of abstinence.

One way around the difficulties posed by the goal of abstinence is to consider it as a future goal towards which the person is working in stages. In this case, improvement itself is rewarded. Under this system, reducing consumption from one quart of vodka per day to one pint per day is considered a worthwhile gain, not a failure. Rewarding improvement, however, does not mean accepting continued drinking as an end. Although the goal may still be abstinence, this system recognizes that not everyone is able to get there immediately.

In a small minority of cases, controlled drinking is a reasonable goal. For most people, however, it is not a useful long-term aim. But even abstinence alone cannot be the whole goal of treatment. Because alcoholism is a way of life, in most cases, other changes must take place for treatment to be successful. Eliminating only the drinking behavior disrupts a person's life without necessarily making any positive improvements in it. Thus, good treatment must be designed to help the recovering alcoholic deal with the world in new and more successful ways. These people must develop new social skills and generally restructure their living patterns. Approximately half of all recovering alcoholics have other personality disorders that preceded, but did not necessarily cause, their alcoholism. These other problems, too, require attention.

Alcoholism is not the same for every alcoholic. There are common elements, but patterns of drinking, associated physical problems, and the particular life circumstances of each person make it increasingly clear that a narrow, uniform approach to treatment will not suffice.

In contrast to even 15 years ago, the availability of treatment services today is enormous. In the past an individual might have had to travel a great distance to get to the nearest alcoholism treatment program, but today it seems that almost every county, community, or neighboring community provides services.

The treatment of alcoholism is still in a transition period. The trend, however, is to recognize the unique qualities of each individual alcoholic and, based on his or her particular life circumstances, make a careful selection of the proper treatment plan. Although it can be said that treatment of the individual alcoholic is currently replacing general treatment

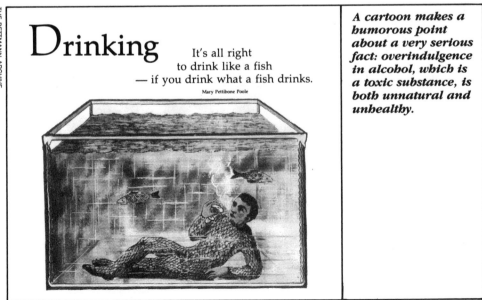

THE BETTMANN ARCHIVE

Drinking

It's all right
to drink like a fish
— if you drink what a fish drinks.

Mary Pettibone Poole

A cartoon makes a humorous point about a very serious fact: overindulgence in alcohol, which is a toxic substance, is both unnatural and unhealthy.

of alcoholism, treatment does still fall into certain broad categories and generally proceeds in discernible stages. The following sections discuss some of the key modern treatment practices and describe how they work.

Reaching the Alcoholic

Because denial is such a major part of most alcoholic behavior, one of the most difficult parts of treatment is just to get the alcoholic to begin the process. Denial is a normal defense mechanism. When used appropriately, it separates people from their potentially immobilizing problems and allows them to take risks and move forward in life. But denial can also sometimes keep a person from evaluating his or her true circumstances.

Problem drinkers and alcoholics often deny they have a problem for several reasons. First, they perceive alcohol as a friend that has helped reduce anxiety, ease pain, express or inhibit feelings, and aid interactions with other people. Thus, to accept alcohol as a source of difficulty is to deny its many rewards. Second, to acknowledge that one does not have the self-control to limit or cease drinking is to admit to a depen-

Homeless men share a bottle of cheap wine on Denver's skid row. Alcoholism is more than a drinking problem or disease — it can become a way of life.

AP/WIDE WORLD PHOTOS

dency on a substance that most other people can handle. Admission of this could make the alcohol abuser feel inferior to others. Third, to acknowledge a lack of self-control implies a need for treatment. For many people who have an exaggerated need to be independent, the idea of requiring help from others, of not being able to handle the problem, is a sign of weakness. And fourth, because the person has a negative stereotype of an alcoholic, he or she can not bear to be considered a member of this group.

Even after alcoholics take the step and enter treatment, they often continue to deny that they have a problem or, if they do admit it, they believe they can handle it by themselves. A major task for the alcoholism treatment specialist, therefore, is to break through this wall of denial. Although this is very difficult, several strategies help. One is to work with the person to draw up a list of the negative consequences of his or her drinking, such as health problems, interpersonal difficulties, family problems, financial crises, and deteriorating job performance. The person reviews the list in one sitting, and this experience may finally make him or her see the reality and enormity of the drinking problem.

Alcoholics at a rehabilitation center attend an AA "step" meeting focusing upon the fifth tenet of the group's 12-step program.

AP/WIDE WORLD PHOTOS

If the first tactic fails, the drinker can be confronted with significant people in his or her life who entreat the person to recognize the problem. An employed person can be threatened with job loss if he or she does not take steps to correct the drinking habits that are interfering with satisfactory work performance.

Why would a problem drinker or alcoholic continue to turn a deaf ear to the pleas of health professionals and concerned family members, friends, and employers? Usually it is because the person strongly resists the idea that he or she will have to stop drinking, a recommendation put forth by virtually all treatment specialists. The person wants to be able to drink just like everyone else drinks. He or she may feel that the need to refuse drinks will be a telltale sign of an alcohol problem. But most important is the fact that the problem drinker is often desperately afraid to give up the use of alcohol because he or she is not sure that life can go on without it. Often the person does not verbalize this fear, but again chooses to deny the dependency on this drug.

There is still another popular approach to breaking through denial. It consists of convincing the person that al-

Botticelli's **The Derelict** *evokes the despair that often accompanies addiction. Yet many problem drinkers resist getting help because they fear the prospect of life without alcohol more than they fear the ravages of the disease itself.*

coholism is a disease that cannot be controlled but must be treated. Why might this approach work? The confronted problem drinker often blames his or her need to drink on outside pressures imposed by other people. At another time, the same drinker may blame him- or herself, not necessarily for the out-of-control drinking, but for the inability to handle life stress. The disease concept can break this cycle of recrimination. The problem drinker is told that no one is to blame, and that because of the disease his or her body has become unable to metabolize alcohol normally. In addition, in many instances the person can be warned that alcohol is beginning to harm various organs of the body, and that continued use may cause permanent and possibly fatal damage. Frequently this approach is tried first, often with good results.

Sometimes all of the methods are used, either one at a time or simultaneously. Usually it is done by the treatment specialist, but occasionally by a lay person who is trying to get the problem drinker to contact a professional. The breakthrough, however, may take days, weeks, months, or even years, and the person's attempts to face reality may alternate with periods of problem drinking.

Stages of Treatment

Successful treatment of alcoholism requires several different stages. These stages can include detoxification, rehabilitation, and outpatient therapy.

Detoxification. Whenever alcohol is present in the body, the liver works to remove it from the bloodstream and detoxify the body. Since problem drinkers consume alcohol faster than their livers can remove it, excess alcohol circulates freely throughout the body and inflicts its toxic effects. Cessation of drinking allows the liver to catch up on its work of metabolizing the alcohol. For those whose bodies have grown accustomed to the presence of large amounts of alcohol, this process happens more rapidly than the time it takes the body's cells to become used to functioning in the absence of alcohol. In response to this withdrawal, or the relatively rapid drop of blood-alcohol level, such symptoms as sweating, nervous tremors, rapid heart rate, and increased blood pressure appear. A hangover, which is really a kind of "mini-withdrawal," is far less severe.

One technique for treating withdrawal is to hospitalize the drinker for about five days. The use of inpatient detoxification units is a relatively new concept, having begun in the 1970s. During the treatment, a tranquilizer, which serves to ease discomfort, is administered. Medical personnel are alert to the possibility that the patient will experience a seizure (a symptom of the early stage of detoxification) or the more serious *delirium tremens*, or DTs, which are characterized by confusion, disorientation, hallucinations, and intense agitation. If untreated, the DTs are fatal in as many as 20% of the cases.

Recent research has shown that nonhospital detoxification in "sobering-up stations" is also usually effective. *Ambulatory detoxification*, a less expensive procedure that allows the drinker to sleep at home but to visit the hospital clinic each day for examination and/or medication, also works for some people. Hospital detoxification, however, does have the advantage of allowing a treatment worker to prepare the often reluctant alcoholic for the next stage of treatment.

Actor Don Johnson (left; shown here with his "Miami Vice" TV series co-star, Philip Michael Thomas) has made public his recovery from a serious drinking problem in the hope that his experience will motivate and inspire other alcoholics to seek treatment.

AP/WIDE WORLD PHOTOS

Rehabilitation. This procedure, whose goal is to help the alcoholic regain productivity, usually follows detoxification and can last from four weeks to six months. The concept of following detoxification with comprehensive care is very new. With the early success of AA, many people believed that after detoxification all that an alcoholic needed was to begin attending AA meetings. Other intervention, especially from professionals, was considered unnecessary and even harmful.

Some people still take this position, but with the influx of health professionals into the alcoholism field, treatments used elsewhere in the mental-health field are now available to recovering alcoholics. These supplemental treatments have proven to be helpful in many cases, and today it is widely accepted that the combination of these therapies and AA attendance offers people the best chance for recovery. At the outset, these traditional therapies were not entirely suited to the specific problems of alcoholics, a fact that further engendered doubts about their usefulness. However, in the last 12 years these therapies have been adapted to be more applicable to the treatment of alcohol abusers.

Inpatient residential treatment may take place in a hospital, but more often it occurs in free-standing, privately

AP/WIDE WORLD PHOTOS

A nurses' station at a busy Los Angeles hospital. One technique for treating alcohol withdrawal is to hospitalize the drinker for about five days. The use of inpatient detoxification units involves the use of tranquilizers, which ease discomfort.

owned alcoholism rehabilitation centers, more and more of which are being set up across the United States. An extended stay in such a facility provides the alcoholic with a "time out." The alcoholic lives in an environment that prohibits the use of alcohol or any other nonprescribed drugs. Individual and group counseling, alcohol education, and the relearning of social skills are provided. Such a program offers an alcoholic a good headstart on sobriety as he or she starts learning how to cope without alcohol.

An increasingly popular alternative to the inpatient residential stay is intensive outpatient rehabilitation. Available at a fraction of the cost of an inpatient program, it provides an opportunity for the less impaired problem drinker or alcoholic identified through early intervention at the work site to obtain treatment without having to leave the job, family, and community. Less artificial than inpatient programs and not overprotective and isolated, intensive outpatient rehabilitation, which usually requires participation four nights a week for twelve weeks, enables the client to deal with conflicts, anxieties, and problems as they exist in the real world on a day-to-day basis.

A halfway house for prisoners with drug problems. Facilities such as this offer recovering addicts the chance to develop personal relationships, acquire social skills, and learn the meaning of individual responsibility.

Some patients in rehabilitation, however, cannot really relearn skills, because they never acquired them. For them, the process is *habilitation*, rather than rehabilitation. In general, people who are relearning skills they once had are more likely to be productive after treatment than those individuals who are learning them for the first time. Alcoholics in need of habilitation may have few skills and few outside resources, such as family, friends, home, or employment, to support them through sobriety. Because of this, they are sometimes referred to *halfway houses*. These facilities are even more supervised than the usual residential settings and offer opportunities for supervised communal living. Here the alcoholic develops personal relationships, acquires social and coping skills, and learns about individual responsibility. The halfway house experience can last from three months to two years.

Outpatient Therapy. When the recovering alcoholic seems ready, he or she is almost always encouraged to move to this stage of treatment. In recent years it has become common for problem drinkers who seek help but who do not require detoxification to enter an outpatient program

Lou Solon, a recovered alcoholic, and Phyllis Phillips, an employment agency president, work together in hiring former alcoholics. Vocational training is an important aspect of many programs designed to help alcoholics resume productive lives.

directly. Because recovery is a long-term process, outpatient therapy can be an open-ended experience.

Outpatient clinics may be public or private, hospital-based or free-standing. They offer a variety of treatment practices, including individual, group, and family counseling and frequently onsite meetings of AA. A wide range of additional therapies are more or less available depending on the treatment philosophy, the nature of the population served, and the budget. The most common therapies are meditation, relaxation, acupuncture, biofeedback, nutritional counseling, assertiveness training, art therapy, writing therapy, dance therapy, occupational therapy, vocational therapy, and recreational therapy. Alcohol education and education classes leading to a high school diploma are also provided. Some treatment workers administer a drug called Antabuse, which causes severe discomfort and illness in people who drink alcohol. Because of this wide variety of treatment therapies, recovering alcoholics can search for the clinic that best suits their needs.

Relapse prevention is a general goal of all alcoholism treatment programs. However, only recently have specific

A group therapy session at a Manhattan Veterans Administration hospital. The task of overcoming addiction is an enormously difficult one; support groups such as these can ease the process.

behavioral prevention techniques — ways of training a person to learn new ways to act within a new lifestyle — been developed to help recovering alcoholics maintain sobriety.

After-Care: Early and Late Sobriety

Few written descriptions can accurately describe the difficulties alcoholics meet as they try to stop drinking. Perhaps at the center of the issue is that alcoholics are terrified of the consequences. They find it especially painful to accept the fact that alcohol, which they had regarded as a "best friend," is now their worst enemy.

Consequently, the alcoholic rarely just gives up drinking; instead, he or she is usually in a constant struggle over this major life change. The denial of dependency, the unwillingness to accept the diagnosis of alcoholism and the prescribed abstinence, the desire to be normal, the craving for alcohol, the belief that one can drink with complete control — all of these make the period known as *early sobriety* a most precarious one.

Many destitute and homeless alcoholics lack family and friends to support their efforts to achieve sobriety. These people may find a home and support system in halfway houses, which provide supervised communal living and caring staff.

Adequate continuing care during early sobriety is crucial. To maintain complete abstinence, recovering alcoholics need support and encouragement. According to many specialists, during this period it is also important that there is an increase in self-esteem and a range of positive experiences in coping with the surroundings, which to the recovering alcoholic may sometimes seem overwhelming.

In *later sobriety*, continuing care helps maintain sobriety. In addition, during this stage, treatment focuses on developing job skills, obtaining further education, and reintegrating the recovering alcoholic into his or her family. They are taught how to socialize and use leisure time productively without anxiety and the use of alcohol. Most important, perhaps, recovering alcoholics are taught how to be assertive, forthright, and clear-thinking without being aggressive or needlessly combative. This entails being able to express positive and negative feelings about others and to others. It also means being able to handle feelings such as anger, anxiety, frustration, guilt, depression, and shame without loss of self-esteem or a return to drinking.

Modes of Treatment

The overwhelming majority of treatment programs offer alcoholism education sessions to their clients. These sessions are necessary for several reasons. First, a person well-informed about his or her disease, disorder, or condition is a more cooperative and involved patient. Second, professionals think patients should understand the philosophy of alcoholism followed by the clinic in which they are being treated. And third, because denial of one's alcoholism is, in part, due to a belief in false stereotypes about alcoholics and alcoholism, a good treatment program enlightens its patients about the true nature of the condition and in this way begins to break through denial and prepare the alcoholic to accept help from others.

Social-Skill Training.

A number of programs help patients develop skills for improved social interactions in the outside world. These may include learning how to be assertive and to express emotions appropriately, to write letters and fill out job applications, and to handle oneself in an interview. Some patients may have these skills, but often they were only able to use them under the influence of alcohol. In treatment, the person is trained to use these skills without the use of alcohol, which means overcoming anxiety, fear of failure, and other negative emotions.

Individual Therapy. Also known as one-on-one therapy, this method gives patients the opportunity to develop a relationship based on confidentiality and unconditional acceptance. Individual therapy can consist of alcoholism counseling, or it can involve psychotherapy, a kind of treatment that explores deep or unconscious feelings. Most people find that initially this type of therapy is difficult. They hesitate to disclose facts about their personal lives, especially when they believe they will be judged. The process is even more difficult for those people who have gone through a series of painful life experiences that have resulted in their not being able to trust others.

Group Therapy. This form of treatment provides an opportunity for social interaction. Although an alcoholism education class could be considered group therapy because it enables patients to realize that their alcohol-related experi-

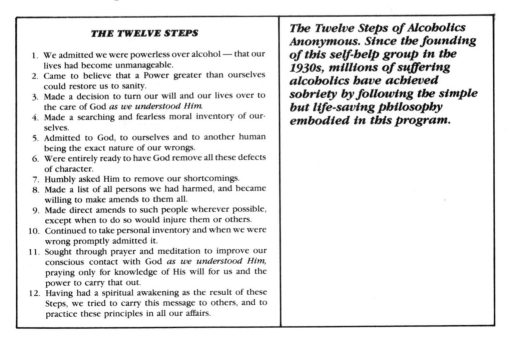

THE TWELVE STEPS

1. We admitted we were powerless over alcohol — that our lives had become unmanageable.
2. Came to believe that a Power greater than ourselves could restore us to sanity.
3. Made a decision to turn our will and our lives over to the care of God *as we understood Him.*
4. Made a searching and fearless moral inventory of ourselves.
5. Admitted to God, to ourselves and to another human being the exact nature of our wrongs.
6. Were entirely ready to have God remove all these defects of character.
7. Humbly asked Him to remove our shortcomings.
8. Made a list of all persons we had harmed, and became willing to make amends to them all.
9. Made direct amends to such people wherever possible, except when to do so would injure them or others.
10. Continued to take personal inventory and when we were wrong promptly admitted it.
11. Sought through prayer and meditation to improve our conscious contact with God *as we understood Him,* praying only for knowledge of His will for us and the power to carry that out.
12. Having had a spiritual awakening as the result of these Steps, we tried to carry this message to others, and to practice these principles in all our affairs.

The Twelve Steps of Alcoholics Anonymous. Since the founding of this self-help group in the 1930s, millions of suffering alcoholics have achieved sobriety by following the simple but life-saving philosophy embodied in this program.

ences are not unique, these classes are not set up to explore deeper psychological problems. More representative types of group therapy use counseling, problem solving, and in-depth therapy groups. In group situations, a participant will react to and interact with other group members in ways similar to those in which he or she interacts with people ouside the group. In a sense, the group experience is a laboratory in which individuals learn how to relate to people and improve relationships.

In addition to providing the potential for social inter-action, which is lacking in individual therapy, group therapy allows an individual to pace his or her own degree of par-ticipation. One disadvantage, however, is the greater stress of group interaction. A group member must relate not only to a single therapist also but to a group of peers. While the group situation can undoubtedly be anxiety-provoking, in the hands of a qualified leader, individuals can benefit from ex-periencing and then overcoming this anxiety. Surviving such an experience can enable group members to see that they have the strength to survive without alcohol.

Marital and Family Counseling. These types of treat-ment give people the opportunity to resolve interpersonal problems. Marital therapy is usually restricted to the husband and wife. Couples therapy is a group experience with four or five pairs, each of which may or may not consist of married individuals. Family therapy involves parents, or a single par-ent, and some or all of the children. Multiple family therapy permits families to observe how other families interact.

Self-Help Groups. Alcoholics Anonymous is the original and prototypical self-help group. In the field of alcoholism, self-help groups initially developed because health-care professionals were not interested in the problems and/or they were having trouble helping alcoholics to maintain sobriety. On a deeper level, it has been suggested that the self-help movement is actually a peer (sibling) approach to handling problems for which professionals (parents) are not available. If this is true, self-help groups represent a symbolic rejection of authority figures.

Paradoxically, in rejecting parental authority, self-help-ers create their own. AA's Twelve Steps, and the rules of offshoot groups such as Narcotics Anonymous, Pills Anonymous, and Families Anonymous, are demanding and require great personal introspection. Members are told that infractions threaten the whole group (the "family"). This suggests that successful social systems require order, whether imposed or self-imposed.

How Effective Is Treatment?

Alcoholism is a stubborn and complex condition, but it is treatable. To quote the National Council on Alcoholism, "Alcoholism is a treatable and beatable disease." Thousands of recovering alcoholics are living proof of this statement. However, most alcoholics who are treated fail to achieve extended sobriety free of periodic relapses. In addition, only a small fraction of those who do attain prolonged sobriety have a fulfilling and productive recovery. Some studies show that alcoholics participating in certain treatment programs do no better than other alcoholics offered only educational information. On average, studies suggest that success rates are no better than 20%.

One must realize that treatment programs have different ways of reporting successes, and, as discussed earlier, success can mean different things to different people. For example, employee assistance programs may define success primarily in terms of improved productivity, though a therapist in a clinic may look for a more complete change in behavior patterns. Similarly, there is disagreement on whether a relapse into alcoholism after a long period of abstinence constitutes total failure — or a partial success. Still, even allowing for all these differences of opinion, it is generally acknowledged that treatment programs are not as successful as one might wish.

Why is this true? Is there something about the addictive process that is unconquerable? Are there absolute limits on what treatment can accomplish?

There is no convincing evidence to suggest that there is something about alcoholism or addiction in general that makes future improvements impossible. However, there do seem to be a number of factors that have worked against successful treatment.

First, alcoholics face psychological difficulties, such as low self-esteem and pervasive feelings of failure. This makes it difficult for most of them to accept the successful outcome of treatment. They may be unconsciously seeking to undo any progress they may have made.

A second, related explanation is that at some level the alcoholism-treatment community (planners, administrators, treatment staff) identifies with the failure-oriented alcoholic and unconsciously creates approaches that result in short-term success and long-term failure. Treatment staff may "catch the disease" from overexposure to alcoholic clients and become demoralized by the pervasive sense of hopelessness that alcoholics often display. For some professionals, overexposure to alcoholic clients manifests itself as "burnout," a sudden lack of motivation and/or a belief on the part of these professionals that they can no longer be useful to their clients.

It has been suggested that this feeling of hopelessness is played out even at the planning and administrative levels of treatment programs. Indeed, some experts feel that the entire alcoholism community, having adopted a philosophy that reflects the hopelessness of the alcoholic, has programmed itself to fail. There is evidence that treatment is more successful when it goes beyond the traditional modes of therapy to include newer kinds of therapies. Despite this evidence, many

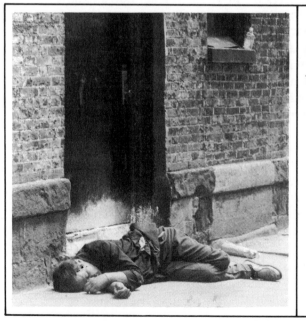

Alcoholics attempting to achieve sobriety often face psychological obstacles such as low self-esteem and feelings of failure. These negative feelings can weaken the alcoholic's resolution to stay sober and increase the likelihood of a relapse.

MICHOS TZOVARAS/ART RESOURCE

new programs follow the same old patterns and ignore the newer and potentially more successful approaches. There is evidence to suggest that this may be because people who operate new programs do not want to risk trying methods that are too different, fearing they may upset the status quo and fail to win acceptance.

Critics of traditional approaches to treating alcoholism claim that such approaches perpetuate certain destructive myths about alcohol. Developed by treatment workers to motivate the recovering alcoholic to remain abstinent, at least three of these myths may have the opposite effect.

Once an alcoholic, always an alcoholic. The purpose of this statement is to inform recovered alcoholics that they always have the potential for relapse. Although this may help the person maintain abstinence, because of alcoholism's stigma it may also lower his or her sense of self-worth. These negative feelings, in turn, can weaken the person's resolve to carry on the attempt to stay sober.

Even one drink will lead to uncontrollable drinking. This is in essence a scare tactic designed to get people to avoid returning to alcohol on even a single occasion. Research indicates, however, that even when alcohol is freely available, one drink does not set up a chain of events inevitably leading to loss of control. It has been hypothesized that if individuals believe uncontrolled drinking is an inevitable consequence of one drink, then those people who do slip may feel further efforts are fruitless. With a more flexible and positive attitude, the alcoholic need not regard slips as fatal.

When an alcoholic returns to drinking after a period of sobriety, the disease process and the level of drinking will manifest themselves at the same level that they would have operated if drinking had continued during the period of sobriety. This assumption is at best a questionable one. It is not even clear that the patterns would have worsened over this period. In fact, the alcoholic who believes this statement and returns to drinking comes to see his or her past efforts at maintaining sobriety as being a waste. This attitude gives the alcoholic very little encouragement to return quickly to abstinence.

ALCOHOLISM CAN TEAR A FAMILY APART

AL-ANON/ ALATEEN CAN HELP
HELP FOR FAMILIES AND FRIENDS OF PROBLEM DRINKERS

As this Al-Anon/Alateen poster shows, alcoholism is a disease that affects every member of the family. Al-Anon and Alateen, offshoots of Alcoholics Anonymous, both began in the 1950s for the purpose of treating spouses, children, and friends of alcoholics.

CHAPTER 7

TREATMENT OF THE FAMILY

*I*t is often said that the modern era of concern about alcoholism began with the creation of Alcoholics Anonymous, in the 1930s. It was not until the 1970s, however, that the role of the family began to be emphasized in discussions about alcoholism. At that time, literature began to appear noting that the average alcoholic adversely affects the lives of at least four other people, most of whom are family members — the often overlooked victims of alcoholism. In effect, then, not only 10–13 million alcoholics but also some 40 million others — nearly 20% of the entire population of the United States — suffer from the ravages of alcoholism.

Given this rather startling statistic, it is no wonder that the treatment of the alcoholic has increasingly focused on the whole family. The creation of Al-Anon (the sister self-help group of AA designed to assist family and friends of alcoholics) was a recognition of the importance of the family in treatment. In groups such as Al-Anon, participants learn how to interact with an alcoholic without exacerbating the problem and how to extricate their lives from the alcoholic's.

Impact of Alcoholism on the Family

If an individual's drinking problem develops suddenly and escalates rapidly, his or her spouse may show little tolerance for it — demanding change and urging treatment. If the drinker's response is unsatisfactory, the marriage is not likely to last.

More often, however, the drinking problem escalates gradually. In this case, the spouse is less likely to object as quickly or firmly, and slowly but surely drinking comes to play a more important role in the relationship. Drinking begins to influence verbal, sexual, and social interactions. Social embarrassment about drinking may reduce contact with fam-

A child drinking from a beer keg. Adults serve as role models for their children. This may partially account for the fact that more than one-fourth of those whose parents are alcoholics eventually develop drinking problems themselves.

UPI/BETTMANN NEWSPHOTOS

ily and friends. Tension increases, and sometimes the non-alcoholic spouse seeks a separate social life.

Through misguided support, the nonalcoholic spouse often unknowingly contributes to the problem by allowing the drinking to continue, and even helping to create an environment in which the drinker is likely to resist seeking help. Despite the tension, the nonalcoholic partner may join the drinker in minimizing the severity of the problem. He or she may even make excuses for the drinker. Many spouses of alcoholics have an opposite-sex alcoholic parent. Some experts speculate that in the current marriage, the spouse is unconsciously trying to replay the parent's marriage. The spouse is trying to change the outcome, however, and save his or her partner from alcoholism. Other spouses seem to want to be martyrs and use the problem drinking as a way of playing out that role.

Many of these marriages end in divorce. The nonalcoholic spouse, no longer able to tolerate what has become a hopeless and unrewarding situation, seeks freedom as a last resort. There are couples who do ride out the storm. They overcome active drinking and begin a journey of recovery during which the marriage is rebuilt and strengthened. On the other hand, it is not uncommon that, after the period of active drinking is over, either the alcoholic or the spouse recognizes that he or she needed the other person only as long as the drinking and the stress was present. In this case, some aspect of the stress-filled period was fulfilling a psychological need of one or both of the partners.

Ending a marriage for these reasons should not necessarily be seen as a failure. For some people, leaving a marriage under these conditions is a sign of personal growth. Once they think the alcohol-abuse problem is over, they come to feel that this particular marriage did not allow them to change. Confronted with the opportunity for liberation, they seek out a new lifestyle.

Children. It may be that the victims of alcoholism who suffer most — perhaps even more than the alcoholic — are the children in the alcoholic home. Although burdened by physical discomfort and deterioration, the adult alcoholic has a lifetime of experience to assist in coping with stresses. Children of alcoholics are more defenseless. Even worse, they are caught in a conflict of unusual intensity. The fighting between the parents in an alcoholic home requires that children take sides or, at best, be ambivalent toward both parents. Frequently, children become go-betweens and conclude, however erroneously, that they themselves are to blame for the problem. They question their own self-worth and become distrustful of adults. Children who have brothers and sisters will often turn only to their siblings for support. Ashamed of what happens at home and fearful that outsiders may find out, children shy away from normal peer relationships. It is not uncommon for children of alcoholics to gravitate toward one another, creating a network devoid of healthy, functioning models with which they can identify.

These children may find that parental fighting is harder to deal with than parental drinking. This is because the alcoholic parent may be more loving, supportive, and interested when intoxicated than he or she is when sober. In general, however, the child is exposed to inconsistent behavior and deprived of support and love. Child abuse and neglect are common in those families headed by an alcoholic parent.

No one can predict exactly how these children will turn out. In general, many children of alcoholics grow up to experience the world in a negative, or at best an ambiguous, way. They tend to fear rejection, lack basic trust, avoid intimacy, have poor interpersonal relationships, and exhibit low tolerance for frustration and poor control of their impulses. Several variables seem to play a role in determining the outcome, including the number of brothers and sisters, the siblings' ages, and the child's place in the birth order; when and how long the alcoholic was active, in treatment, and in recovery; and the availability of the nonalcoholic spouse, members of the extended family, and other resource people such as friends, parents of friends, clergy, and teachers. An additional significant factor is the child's constitution or innate strengths and abilities to withstand problems.

Between one-fourth and one-third of the children of alcoholics become alcoholics themselves. Few people seem to care about what happens to the majority of adult children of alcoholics who do not become alcoholics. The presumption is that if they have eluded the grasp of alcoholism, then they must not need help. This is often very far from the truth. In fact, these individuals often bring into adulthood many of the personal problems they acquired during childhood. Life does go on, but it is not all that they wish it to be. As adults, some of them become alcoholics; others actively avoid alcohol and/ or intimacy with potentially alcoholic spouses or friends. Some of these adults may live in near-normal fashion, though they continue to be affected in varying degrees by relatively minor but unresolved alcohol-related conflicts.

Like their alcoholic parents, adult children of alcoholics tend to minimize the way in which alcohol contributes to their life difficulties. Although many recognize and even discuss the effect that living with an alcoholic parent had on their lives, it is common for them to fail to acknowledge fully the strength of the impact. In seeking professional mental-health services for one kind of problem or another, they often offhandedly mention that they had an alcoholic parent. Even more common, therapists learn of their clients' past while

This poster was part of a campaign waged by the French government to combat alcohol abuse. France has one of the highest rates of alcoholism in the world.

taking routine histories, but fail to see its significance because they have little knowledge of alcoholism. Thanks to a group of alcoholism treatment and alcoholism prevention specialists who are themselves children of alcoholics, society has recently become more aware of this segment of the population, and psychotherapists have begun to recognize the full importance of this early life experience.

In all too many cases, children who kept the fact of family alcoholism a secret grow up to be adults who hide this fact from themselves. It is understandable that these people would wish to avoid the painful memories that would result from a reexamination of their childhoods. But unless they work through this part of their lives, they will never resolve the conflicts that prevent them from living fully.

Family Therapy for Alcoholism

Family therapy is the application of therapeutic techniques to a family to bring about constructive changes in the behavior of its members. In this context, "family" is defined as

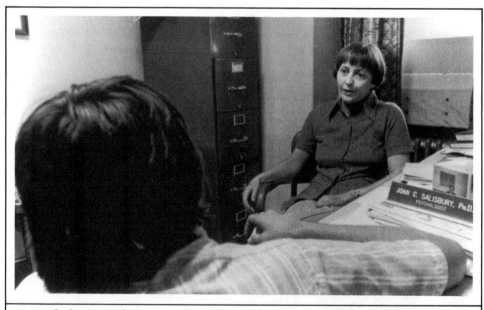

A psychologist with a young patient in an alcohol treatment program. Many of the children of alcoholics who do not themselves develop drinking problems nonetheless carry lifelong emotional scars.

those members living in the same household — regardless of whether they comprise two, three, or four generations — and including other family members living outside the home who have a significant impact on the lives of the group's members.

One reason for using family therapy to deal with alcoholism is that to treat only the alcoholics is to treat them out of context of their usual living environment. In other words, the structure and function of a family may influence the development of alcoholism and the process of recovery. A second reason to use family therapy is that this approach assumes that the alcoholic is not necessarily the only person who needs help. As the formation of Al-Anon suggests, frequently other people in the family also require treatment. Family therapy brings the members of a family together to face and deal with alcohol-related issues. In effect, the patient is the whole family unit.

Although the family of an alcoholic has definite problems, family members often acquire behavior patterns that are, in part, an adaptation to living with the alcoholic. The family as a functional unit attains a kind of *steady state*, or equilibrium, the existence of which allows family members to continue to function. What family therapy tries to do is get the family members to recognize what they are doing, upset this steady state, and then establish a new living situation based on a set of healthier, more productive behaviors.

Troubled adolescents discuss their problems at a hospital therapy session. Young people frequently turn to alcohol in a futile attempt to relieve emotional conflicts and ease social anxieties.

Decisions regarding drinking habits are often made during adolescence. Ultimately, it is up to the individual to weigh the immediate pleasures of drinking against its many potential dangers.

CHAPTER 8

WHAT CAN YOU DO?

For as long as alcoholic beverages have been in existence, they have brought people both pleasure and problems. In contemporary society in the United States — and throughout the world — there is a growing awareness of just how powerful an effect alcohol can have on people's lives. The acute and chronic use of alcohol can have a devastating impact on all aspects — physical, psychological, and social — of life.

After reading and learning about the facts and theories of alcohol and alcoholism, an important question remains: How will you handle alcohol and drinking in the future? You must decide if you will drink at all, and, if so, how you will drink.

Remember what drinking can do. First, if you are a new drinker, it can give you a false sense of having mastered a social skill, of having become an adult. Second, and more dangerous, drinking can create a feeling of temporary well-being that can give the drinker the illusion of being invulnerable. Everyone likes to feel good, and everyone welcomes a sense of personal power and competence. This is one reason drinking is attractive to people of all ages and in many different circumstances. However, the temptation to acquire these feelings primarily through alcohol is especially strong in adolescents.

During the teenage years you must decide who you are. You must develop and strengthen your sense of self-esteem.

This task is not easy. This is a highly competitive society, one in which people are all too apt to measure themselves against others, find themselves lacking, and feel uncomfortable with who and what they are. It can take many years for people to accept their failings and come to value themselves regardless of what others think. Some people never gain this confidence in themselves at all. It is not surprising, then, that in the teenage years you may occasionally — or even often — feel weighed down by the challenge of building self-esteem.

Another related and equally difficult challenge of the teenage years is the three-way conflict between teenagers, their parents, and their peers. All three parties have an idea of how teenagers should act, what they should value, and what they should become. Because in this society there are so many choices about how to live and act, there is a good chance that these ideas will conflict with one another. Adolescence is the time to try to resolve these differences, but in their attempt to do so, teenagers often find that neither their peers nor their parents seem to understand their needs and aspirations.

Sometimes teenagers decide that the only way to resolve these conflicts and challenges is to resort to alcohol, which, because it produces temporary feelings of release, competence, and power, seems like an easy answer. But, in the end, drinking can not bring about true self-esteem, nor can it resolve conflicts. Instead, as a drinker continues to abuse alcohol in the vain hope that it will solve painful issues of personal growth, he or she can become addicted to alcohol.

Merely understanding the facts about alcohol is not the same as reaching decisions about personal behavior. You must be willing to apply what you have learned; you must try to become more conscious of why you act as you do. In the case of alcohol, this means that you need to examine your own drinking behavior, decide if you are using alcohol to escape difficult personal issues, and then judge if this is the healthiest and most productive way to behave.

Ten or twenty years ago, drinking among teenagers was often regarded as a not too serious problem, even when it resulted in damage to property or injury to another individual. A "slap on the wrist" was often the extent of the punishment. Today, however, society is much less willing to overlook the consequences of alcohol abuse. The trend is for society to assign the drinkers the responsibility for actions made under the influence of alcohol, even if they can not remember what they did.

Holding drinkers accountable for their actions seems to be a healthy trend. Part of personal growth is learning that you must take responsibility for your behavior. In the fullest sense, this means making a greater effort to understand why you act as you do, which necessarily includes facing difficult conflicts about your identity. Without this understanding, it is impossible to take real control of your life.

A substance such as alcohol has two sides. If used intelligently and thoughtfully, it may be relatively harmless. However, if it is used to avoid confronting the conflicts, risks, and responsibilities everyone must face, alcohol can be extremely dangerous. Ultimately, only you can decide if you are in danger of becoming dependent on alcohol. And, if you conclude that you are indeed at risk, only you can take the necessary steps to change.

UPI/BETTMANN NEWSPHOTOS

The rock star Madonna is one of several celebrities who have publicly renounced drugs and alcohol. She says that she does not need any artificial stimulants to give her confidence and energy.

APPENDIX

STATE AGENCIES FOR THE PREVENTION AND TREATMENT OF DRUG ABUSE

ALABAMA

Department of Mental Health
Division of Mental Illness and
  Substance Abuse Community
  Programs
200 Interstate Park Drive
P.O. Box 3710
Montgomery, AL 36193
(205) 271-9253

ALASKA

Department of Health and Social
  Services
Office of Alcoholism and Drug
  Abuse
Pouch H-05-F
Juneau, AK 99811
(907) 586-6201

ARIZONA

Department of Health Services
Division of Behavioral Health
  Services
Bureau of Community Services
Alcohol Abuse and Alcoholism
  Section
2500 East Van Buren
Phoenix, AZ 85008
(602) 255-1238

Department of Health Services
Division of Behavioral Health
  Services
Bureau of Community Services
Drug Abuse Section
2500 East Van Buren
Phoenix, AZ 85008
(602) 255-1240

ARKANSAS

Department of Human Services
Office on Alcohol and Drug Abuse
  Prevention
1515 West 7th Avenue
Suite 310
Little Rock, AR 72202
(501) 371-2603

CALIFORNIA

Department of Alcohol and Drug
  Abuse
111 Capitol Mall
Sacramento, CA 95814
(916) 445-1940

COLORADO

Department of Health
Alcohol and Drug Abuse Division
4210 East 11th Avenue
Denver, CO 80220
(303) 320-6137

CONNECTICUT

Alcohol and Drug Abuse
  Commission
999 Asylum Avenue
3rd Floor
Hartford, CT 06105
(203) 566-4145

DELAWARE

Division of Mental Health
Bureau of Alcoholism and Drug
  Abuse
1901 North Dupont Highway
Newcastle, DE 19720
(302) 421-6101

DISTRICT OF COLUMBIA
Department of Human Services
Office of Health Planning and
    Development
601 Indiana Avenue, NW
Suite 500
Washington, D.C. 20004
(202) 724-5641

FLORIDA
Department of Health and
    Rehabilitative Services
Alcoholic Rehabilitation Program
1317 Winewood Boulevard
Room 187A
Tallahassee, FL 32301
(904) 488-0396

Department of Health and
    Rehabilitative Services
Drug Abuse Program
1317 Winewood Boulevard
Building 6, Room 155
Tallahassee, FL 32301
(904) 488-0900

GEORGIA
Department of Human Resources
Division of Mental Health and
    Mental Retardation
Alcohol and Drug Section
618 Ponce De Leon Avenue, NE
Atlanta, GA 30365-2101
(404) 894-4785

HAWAII
Department of Health
Mental Health Division
Alcohol and Drug Abuse Branch
1250 Punch Bowl Street
P.O. Box 3378
Honolulu, HI 96801
(808) 548-4280

IDAHO
Department of Health and Welfare
Bureau of Preventive Medicine
Substance Abuse Section
450 West State
Boise, ID 83720
(208) 334-4368

ILLINOIS
Department of Mental Health and
    Developmental Disabilities
Division of Alcoholism
160 North La Salle Street
Room 1500
Chicago, IL 60601
(312) 793-2907

Illinois Dangerous Drugs
    Commission
300 North State Street
Suite 1500
Chicago, IL 60610
(312) 822-9860

INDIANA
Department of Mental Health
Division of Addiction Services
429 North Pennsylvania Street
Indianapolis, IN 46204
(317) 232-7816

IOWA
Department of Substance Abuse
505 5th Avenue
Insurance Exchange Building
Suite 202
Des Moines, IA 50319
(515) 281-3641

KANSAS
Department of Social Rehabilitation
Alcohol and Drug Abuse Services
2700 West 6th Street
Biddle Building
Topeka, KS 66606
(913) 296-3925

KENTUCKY
Cabinet for Human Resources
Department of Health Services
Substance Abuse Branch
275 East Main Street
Frankfort, KY 40601
(502) 564-2880

LOUISIANA
Department of Health and Human
  Resources
Office of Mental Health and
  Substance Abuse
655 North 5th Street
P.O. Box 4049
Baton Rouge, LA 70821
(504) 342-2565

MAINE
Department of Human Services
Office of Alcoholism and Drug
  Abuse Prevention
Bureau of Rehabilitation
32 Winthrop Street
Augusta, ME 04330
(207) 289-2781

MARYLAND
Alcoholism Control Administration
201 West Preston Street
Fourth Floor
Baltimore, MD 21201
(301) 383-2977

State Health Department
Drug Abuse Administration
201 West Preston Street
Baltimore, MD 21201
(301) 383-3312

MASSACHUSETTS
Department of Public Health
Division of Alcoholism
755 Boylston Street
Sixth Floor
Boston, MA 02116
(617) 727-1960

Department of Public Health
Division of Drug Rehabilitation
600 Washington Street
Boston, MA 02114
(617) 727-8617

MICHIGAN
Department of Public Health
Office of Substance Abuse Services
3500 North Logan Street
P.O. Box 30035
Lansing, MI 48909
(517) 373-8603

MINNESOTA
Department of Public Welfare
Chemical Dependency Program
  Division
Centennial Building
658 Cedar Street
4th Floor
Saint Paul, MN 55155
(612) 296-4614

MISSISSIPPI
Department of Mental Health
Division of Alcohol and Drug Abuse
1102 Robert E. Lee Building
Jackson, MS 39201
(601) 359-1297

MISSOURI
Department of Mental Health
Division of Alcoholism and Drug
  Abuse
2002 Missouri Boulevard
P.O. Box 687
Jefferson City, MO 65102
(314) 751-4942

MONTANA
Department of Institutions
Alcohol and Drug Abuse Division
1539 11th Avenue
Helena, MT 59620
(406) 449-2827

NEBRASKA
Department of Public Institutions
Division of Alcoholism and Drug Abuse
801 West Van Dorn Street
P.O. Box 94728
Lincoln, NB 68509
(402) 471-2851, Ext. 415

NEVADA
Department of Human Resources
Bureau of Alcohol and Drug Abuse
505 East King Street
Carson City, NV 89710
(702) 885-4790

NEW HAMPSHIRE
Department of Health and Welfare
Office of Alcohol and Drug Abuse
   Prevention
Hazen Drive
Health and Welfare Building
Concord, NH 03301
(603) 271-4627

NEW JERSEY
Department of Health
Division of Alcoholism
129 East Hanover Street CN 362
Trenton, NJ 08625
(609) 292-8949

Department of Health
Division of Narcotic and Drug Abuse
   Control
129 East Hanover Street CN 362
Trenton, NJ 08625
(609) 292-8949

NEW MEXICO
Health and Environment Department
Behavioral Services Division
Substance Abuse Bureau
725 Saint Michaels Drive
P.O. Box 968
Santa Fe, NM 87503
(505) 984-0020, Ext. 304

NEW YORK
Division of Alcoholism and Alcohol
   Abuse
194 Washington Avenue
Albany, NY 12210
(518) 474-5417

Division of Substance Abuse
   Services
Executive Park South
Box 8200
Albany, NY 12203
(518) 457-7629

NORTH CAROLINA
Department of Human Resources
Division of Mental Health, Mental
   Retardation and Substance Abuse
   Services
Alcohol and Drug Abuse Services
325 North Salisbury Street
Albemarle Building
Raleigh, NC 27611
(919) 733-4670

NORTH DAKOTA
Department of Human Services
Division of Alcoholism and Drug
   Abuse
State Capitol Building
Bismarck, ND 58505
(701) 224-2767

OHIO
Department of Health
Division of Alcoholism
246 North High Street
P.O. Box 118
Columbus, OH 43216
(614) 466-3543

Department of Mental Health
Bureau of Drug Abuse
65 South Front Street
Columbus, OH 43215
(614) 466-9023

OKLAHOMA

Department of Mental Health
Alcohol and Drug Programs
4545 North Lincoln Boulevard
Suite 100 East Terrace
P.O. Box 53277
Oklahoma City, OK 73152
(405) 521-0044

OREGON

Department of Human Resources
Mental Health Division
Office of Programs for Alcohol and
  Drug Problems
2575 Bittern Street, NE
Salem, OR 97310
(503) 378-2163

PENNSYLVANIA

Department of Health
Office of Drug and Alcohol
  Programs
Commonwealth and Forster Avenues
Health and Welfare Building
P.O. Box 90
Harrisburg, PA 17108
(717) 787-9857

RHODE ISLAND

Department of Mental Health,
  Mental Retardation and Hospitals
Division of Substance Abuse
Substance Abuse Administration
  Building
Cranston, RI 02920
(401) 464-2091

SOUTH CAROLINA

Commission on Alcohol and Drug
  Abuse
3700 Forest Drive
Columbia, SC 29204
(803) 758-2521

SOUTH DAKOTA

Department of Health
Division of Alcohol and Drug Abuse
523 East Capitol, Joe Foss Building
Pierre, SD 57501
(605) 773-4806

TENNESSEE

Department of Mental Health and
  Mental Retardation
Alcohol and Drug Abuse Services
505 Deaderick Street
James K. Polk Building, Fourth Floor
Nashville, TN 37219
(615) 741-1921

TEXAS

Commission on Alcoholism
809 Sam Houston State Office Building
Austin, TX 78701
(512) 475-2577

Department of Community Affairs
Drug Abuse Prevention Division
2015 South Interstate Highway 35
P.O. Box 13166
Austin, TX 78711
(512) 443-4100

UTAH

Department of Social Services
Division of Alcoholism and Drugs
150 West North Temple
Suite 350
P.O. Box 2500
Salt Lake City, UT 84110
(801) 533-6532

VERMONT

Agency of Human Services
Department of Social and
  Rehabilitation Services
Alcohol and Drug Abuse Division
103 South Main Street
Waterbury, VT 05676
(802) 241-2170

VIRGINIA
Department of Mental Health and
  Mental Retardation
Division of Substance Abuse
109 Governor Street
P.O. Box 1797
Richmond, VA 23214
(804) 786-5313

WASHINGTON
Department of Social and Health
  Service
Bureau of Alcohol and Substance
  Abuse
Office Building—44 W
Olympia, WA 98504
(206) 753-5866

WEST VIRGINIA
Department of Health
Office of Behavioral Health Services
Division on Alcoholism and Drug
  Abuse
1800 Washington Street East
Building 3 Room 451
Charleston, WV 25305
(304) 348-2276

WISCONSIN
Department of Health and Social
  Services
Division of Community Services
Bureau of Community Programs
Alcohol and Other Drug Abuse
  Program Office
1 West Wilson Street
P.O. Box 7851
Madison, WI 53707
(608) 266-2717

WYOMING
Alcohol and Drug Abuse Programs
Hathaway Building
Cheyenne, WY 82002
(307) 777-7115, Ext. 7118

GUAM
Mental Health & Substance Abuse
  Agency
P.O. Box 20999
Guam 96921

PUERTO RICO
Department of Addiction Control
  Services
Alcohol Abuse Programs
P.O. Box B-Y Rio Piedras Station
Rio Piedras, PR 00928
(809) 763-5014

Department of Addiction Control
  Services
Drug Abuse Programs
P.O. Box B-Y Rio Piedras Station
Rio Piedras, PR 00928
(809) 764-8140

VIRGIN ISLANDS
Division of Mental Health,
    Alcoholism & Drug Dependency
    Services
P.O. Box 7329
Saint Thomas, Virgin Islands 00801
(809) 774-7265

AMERICAN SAMOA
LBJ Tropical Medical Center
Department of Mental Health Clinic
Pago Pago, American Samoa 96799

TRUST TERRITORIES
Director of Health Services
Office of the High Commissioner
Saipan, Trust Territories 96950

Further Reading

Armyr, G., Elmer, A. and Herz, U. *Alcohol in the World of the 80s: Habits, Attitudes, Preventive Policies and Voluntary Efforts.* Stockholm: Sober Forlags, 1982.

Cahalan, D., Cisin, I.H. and Crossley, H.M. *American Drinking Practices.* New Brunswick, N J: Rutgers Center of Alcoholic Studies, 1969.

Claypool, Jane. *Alcohol and You.* New York: Franklin Watts, 1981.

Langone, John. *Bombed, Buzzed, Smashed . . . or Sober.* New York: Avon, 1970.

Marshall, M., ed. *Beliefs, Behaviors, and Alcoholic Beverages: A Cross-Cultural Survey.* Ann Arbor, MI: University of Michigan Press, 1979.

Milgram, Gail G. *Coping with Alcohol.* New York: Rosen Group, 1980.

Glossary

anesthetic a drug that produces loss of sensation, sometimes without loss of consciousness

antihistamine a drug that inhibits the action of histamine and thus reduces the allergic response

antiseptic an agent that inhibits the growth of microorganisms and thus prevents infection

caffeine trimethylxanthine; a central nervous system stimulant found in coffee, tea, cocoa, various soft drinks, and often in combination with other drugs to enhance their effects

cholesterol a type of alcohol found in animal tissues and occurring in such substances as egg yolks, fats, and various oils; important in certain metabolic processes

cirrhosis a chronic disease of the liver characterized by the development of scar tissue to replace those parts of the liver damaged by inflammation and other disease-related symptoms; results in decreased function of liver cells and increased resistance to blood flow through the liver

cocaine the primary psychoactive ingredient in the coca plant and a behavioral stimulant

delirium tremens a disorder resulting from withdrawal from chronic alcohol use characterized by confusion, disorientation, hallucinations, and intense agitation

detoxification the process by which an addicted individual is gradually withdrawn from the abused drug, usually under medical supervision and sometimes in conjunction with the administration of other drugs

disorder a pathological condition of the mind or body

drug any substance derived from a plant or synthesized in a laboratory that when ingested, injected, sniffed, inhaled, or absorbed from the skin affects bodily functions

dysfunction inadequate, abnormal, or impaired function of part of the body

euphoria a mental high characterized by a sense of well-being

gastritis inflammation of the stomach

hallucination a sensory impression that has no basis in external stimulation

heroin a semisynthetic opiate produced by a chemical modification of morphine

limbic system an evolutionarily more primitive part of the brain associated with memory and sexual and emotional behavior

metabolism the chemical changes in the living cell by which energy is provided for the vital processes and activities and by which new material is assimilated to repair cell structures; or, the process that uses enzymes to convert one substance into compounds that can be easily eliminated from the body

methadone a synthetic opiate producing effects similar to morphine's effects and used to treat pain associated with terminal cancer and in the treatment of heroin addicts

morphine the principal psychoactive ingredient of opium that produces sleep or a state of stupor, and is used as the standard against which all morphine-like drugs are compared

myocarditis inflammation of the muscles of the heart

neurosis a disorder of the thought processes that does not include a loss of contact with reality

obsessive-compulsive characterized by the inclination to perform rituals repeatedly to relieve anxiety

opiate a compound from the milky juice of the poppy plant *Papaver somniferum*, including opium, morphine, codeine, and their derivatives, such as heroin

paranoia a mental condition characterized by extreme suspiciousness, fear, delusions, and in extreme cases hallucinations

physical dependence an adaptation of the body to the presence of a drug such that its absence produces withdrawal symptoms

psychoactive altering mood and/or behavior

psychological dependence a condition in which the drug user craves a drug to maintain a sense of well-being and feels discomfort when deprived of it

rebound effect an increase in the activity level of the cells following the abrupt cessation of alcohol use

sedative a drug that produces calmness, relaxation, and, at high doses, sleep; includes barbiturates

stimulant any drug that increases brain activity which results in the sensation of greater energy, euphoria, and increased alertness

tolerance a decrease of susceptibility to the effects of a drug due to its continued administration, resulting in the user's need to increase the drug dosage to achieve the effects experienced previously

tranquilizer a drug that has calming, relaxing effects

withdrawal the physiological and psychological effects of discontinued usage of a drug

Index

Ross Fishman, Ph.D., received his degree from the University of Pittsburgh in psychology. He is the former director of Education and Training at the Alcoholism Council of Greater New York. Currently, he is the director of Innovative Health Systems, Inc. in White Plains, New York. He has taught many courses on alcohol and alcoholism and has written on teenage drinking, confidentiality in alcoholism treatment, alcohol and methadone, and employee alcoholism programs.

Solomon H. Snyder, M.D., is Distinguished Service Professor of Neuroscience, Pharmacology and Psychiatry at The Johns Hopkins University School of Medicine. He has served as president of the Society for Neuroscience and in 1978 received the Albert Lasker Award in Medical Research. He has authored *Uses of Marijuana, Madness and the Brain, The Troubled Mind, Biological Aspects of Mental Disorder,* and edited *Perspective in Neuropharmacology: A Tribute to Julius Axelrod.* Professor Snyder was a research associate with Dr. Axelrod at the National Institutes of Health.

Barry L. Jacobs, Ph.D., is currently a professor in the program of neuroscience at Princeton University. Professor Jacobs is author of *Serotonin Neurotransmission and Behavior* and *Hallucinogens: Neurochemical, Behavioral and Clinical Perspectives.* He has written many journal articles in the field of neuroscience and contributed numerous chapters to books on behavior and brain science. He has been a member of several panels of the National Institute of Mental Health.

Jerome H. Jaffe, M.D., formerly professor of psychiatry at the College of Physicians and Surgeons, Columbia University, has been named recently Director of the Addiction Research Center of the National Institute on Drug Abuse. Dr. Jaffe is also a psychopharmacologist and has conducted research on a wide range of addictive drugs and developed treatment programs for addicts. He has acted as Special Consultant to the President on Narcotics and Dangerous Drugs and was the first director of the White House Special Action Office for Drug Abuse Prevention.